Management and Industrial Relations Series

6

New product development in engineering

A comparison of the British and
West German machine tool industries

Management and Industrial Relations Series

Social science research has much to contribute to the better understanding and solution of problems in the field of management and industrial relations. The difficulty, however, is that there is frequently a gap between the researcher and the practitioner who wants to use the research results. This new series is designed to make available to practitioners in the relevant fields the results of the best research which the Economic and Social Research Council (ESRC) has supported in the fields of management and industrial relations. The subjects covered and the style adopted will appeal to managers, trade unionists and administrators because there will be an emphasis upon the practical implications of research findings. But the volumes will also serve as a useful introduction to particular areas for students and teachers of management and industrial relations.

The series is published by Cambridge University Press in collaboration with the Economic and Social Research Council.

Other books in the series

1 *Lost managers: supervisors in industry and society* by JOHN CHILD and BRUCE PARTRIDGE
2 *Tackling discrimination in the workplace: an analysis of sex discrimination in Britain* by BRIAN CHIPLIN and PETER SLOANE
3 *Inflation accounting: an introduction to the debate* by GEOFFREY WHITTINGTON
4 *Industrial relations and management strategy* edited by KEITH THURLEY and STEPHEN WOOD
5 *Power in a trade union: the role of the district committee in the AUEW* by LARRY JAMES

New product development in engineering

A comparison of the British and West German machine tool industries

STEPHEN T. PARKINSON

The right of the University of Cambridge to print and sell all manner of books was granted by Henry VIII in 1534. The University has printed and published continuously since 1584.

CAMBRIDGE UNIVERSITY PRESS

Cambridge
London New York New Rochelle
Melbourne Sydney

Published by the Press Syndicate of the University of Cambridge
The Pitt Building, Trumpington Street, Cambridge CB2 1RP
32 East 57th Street, New York, NY 10022, USA
296 Beaconsfield Parade, Middle Park, Melbourne 3206, Australia

First published 1984

Printed in Great Britain at the University Press, Cambridge

Library of Congress catalogue card number: 83–20855

British Library cataloguing in publication data

Parkinson, Stephen T.
New product development in engineering: a comparison of the British and West German machine tool industries. – (Management and industrial relations series; 6)
1. Machine tools – Trade and manufacture – Great Britain 2. Technological innovation – Great Britain 3. Machine tools – Trade and manufacture – Germany (West) 4. Technological innovation – Germany (West)
I. Title II. Series
338.4′7621902′0941 HD9703.G72

ISBN 0 521 25796 4

To Lynn

Contents

Preface

This book is the outcome of several years' research into the process of new product development and new product adoption in the engineering industries of the United Kingdom and West Germany. The present work developed from earlier research conducted with Professor Michael J. Baker of the University of Strathclyde, Glasgow. This earlier research had focussed on factors influencing the rate of acceptance of industrial innovation and was sponsored by the Social Science Research Council. Much of the original stimulus and encouragement for writing this book came from Michael Baker and I would like to thank him gratefully for his many useful contributions and continuing support.

In August 1978 the Social Science Research Council and the Deutsche Forschungs Gemeinschaft exchange scheme for social scientists funded a three months' research visit to Munich where I was able to collect much of the field data for the German section of the research study. I gratefully acknowledge their financial assistance. Thanks are also due to the IFO – Institut für Wirtschaftsforschung in Munich for their hospitality and assistance during my stay.

I am also grateful to the many managers in the United Kingdom and West Germany who gave up their time willingly to help with this research enquiry. Without their support this research could not have been completed and I hope that they will find something of practical relevance to their own day-to-day problems in this book.

Thanks are also due to several colleagues who have commented on preliminary drafts of this manuscript. In particular, I would like to acknowledge the helpful comments of Dr Axel Johne of the City University, London, and Dr Roy Rothwell of the Science Policy

Research Unit, Sussex University. I found their comments particularly useful in sorting out difficulties in the manuscript, resolving contradictions and organising the material effectively. Preliminary findings from this research were published in *R and D Management* in April 1981 and July 1982, and I am also grateful for individual readers' comments which helped in the final shaping of this manuscript.

Thanks are also due to Mrs June Peffer who undertook the onerous task of deciphering my handwriting, and produced the final manuscript cheerfully and professionally.

Finally, this work would not have been completed without the continuing support of my wife, Lynn, who provided the necessary balance of encouragement and cajoling to complete the research, and translate its findings into the present book.

University of Strathclyde
Glasgow
March 1983

STEPHEN T. PARKINSON

1

Technological innovation and international competitiveness

Introduction

Technological innovation embodied in new products or processes has become an important source of competitive advantage for all advanced industrial countries. In such countries suppliers have had to put an increasing emphasis on improving product quality in order to compete successfully in international markets. Price competition has become increasingly less attractive as suppliers from developing countries have entered the market, supplying relatively standard products with lower manufacturing costs. This development has been particularly pronounced in the mechanical engineering sectors of the economy, although its effects have been felt throughout manufacturing industry.

Despite the obvious importance of successful new product development to economic performance many British companies are not as successful as their overseas competitors in developing new products in line with the needs of their customers. There is a growing literature which points to the failure of British companies on this important dimension, and contrasts their recent experience with that of competitors from other developed countries (notably West Germany).

Recent analyses have shown an association between the attention paid to improvements in product quality and export trading success. It is clear that British companies' traditionally high share of many export markets has declined partly because of a failure to introduce technically advanced products successfully. It is also clear that one of the major reasons for the relatively high level of import penetration experienced in some British markets in recent years has been the superior quality of the imported product.

Several different explanations have been put forward to account for this comparative decline. The most frequently cited is that the environment within which British firms operate is not conducive to technological innovation. Conventional arguments blame the taxation system as discouraging enterprise, or the money market for being too conservative to support new product ideas. It has been asserted that the domestic market is not large enough to support the development costs of new products or processes and that there are too many uncertainties with overseas markets to proceed without a good domestic base. The government has also been accused of not giving sufficient assistance to new technologies by creating the right climate for investment or focussing too much research and development funding on a few large projects such as defence, aerospace or nuclear power. These explanations are all of the external type. They suggest that it is the system which is to blame rather than management in individual companies who work within this environment.

Whilst there is some merit in these explanations a complementary view is that the United Kingdom's performance in technological innovation is also a function of British managements' attitudes and approach to technical innovation. The comparative technical lag which is evident in many sectors of British manufacturing industry may stem as much from deficiencies in the new product development process and customers' conservatism towards new technology as it does from the external influences cited above. This book examines this proposition in a comparative review of the British and West German machine tool industries and their customers.

International Comparisons

Statistics on the relative position of British industries in world markets are difficult to collate and assess. Those which have been compiled suggest an overall pattern of decline in world market shares held by British companies and an overall contraction of activity. An NEDC (National Economic Development Committee) internal report (1981) commented that the statistics which were available suggested that United Kingdom output and output per head had grown much more slowly in almost all industry groups (excluding clothing, and mining and quarrying – a sector which included North Sea Oil) than the average growth experienced by all OECD (Organisation for Economic Cooperation and Development) members.

Where growth was slower than average this by definition meant a loss of market share.

More detailed information in the same report showed that British industry had been slower than German industry to adapt to changing patterns of world demand and seize opportunities. Electrical machinery (including electronics), chemicals and plastic products were cited as industries which had expanded far more rapidly than the same industries in the United Kingdom. Significantly the industries which grew fastest were static or declining in Germany. These industries included clothing, footwear and leather, which are typically low unit value and are now under intense competition from newly industrialising countries.

The comparative decline of British industries is illustrated in even more striking terms by a comment attributed to McIntosh (Phelps-Brown 1977) in a speech in 1976 when he was Director General of the National Economic Development Office (NEDO). He said: 'The most striking statistic of all those relating to the British economy which I know is that in 1953 our share of world exports of manufactures was 23 per cent and Germany's was 19 per cent. Twenty years later, Germany's share had risen to 23 per cent and ours had fallen to under 8 per cent.'

German industry is frequently used as the basis for comparison with the United Kingdom because of basic similarities in opportunities and constraints. Recent studies of individual sectors have pointed to deficiencies in the United Kingdom which may have contributed to this decline. In particular, the failure to develop new technologies and a persistence with price competition have been seen as major contributory factors. A study undertaken for NEDO (1977) commented:

> 'Superior Design' (or 'fitness for purpose') was the reason for purchasing machinery abroad most frequently cited by UK buyers in response to enquiries carried out by NEDO for the EDC [Economic Development Committee] over the last 10 years (1967–1977) . . . the evidence does suggest that there is a need for the UK to increase the input of technology to, and the design excellence of the products of its mechanical engineering industry, if it is to maintain or increase its competitiveness . . . This appears to have been a significant element in the success of West German,

> Swedish and Japanese industries in world markets . . . British industry still places too much emphasis on the price competitiveness of its products, and seriously under-rates the importance of non-price factors such as delivery, design, after sales service, credit facilities and selling expertise.

A British Institute of Management report (1978) expressed the same theme. This report stressed that: 'Unless the industrialised countries, particularly Great Britain, concentrate on the development of high technology products, they will suffer from the low wage competition of developing nations. British companies should specialise in the manufacture of sophisticated higher value products.'

Rothwell's study for several sectors of the mechanical engineering industry provides further illustration (1979). In this study, Rothwell used value per ton of machinery as a crude measure of technical sophistication. He assumed that as value per ton increased, technical sophistication would increase. Controlling for differences in exchange rates, it appeared that there was a positive relationship between high value per ton (technical sophistication) and international trading success in several sectors including agricultural machinery, fork lift trucks, portable power tools, circular knitting machinery and coal mining machinery. Significantly, as the value per ton of British products fell relative to the value per ton of products from overseas competitors, so British companies had lost market share in these sectors.

Rothwell also cited specific examples to support the overall conclusion. For example, he cited British Northrop: ' . . . which was the first firm in Europe to produce the automatic loom in the 1930's, but declined rapidly from the early 1960's onwards due mainly to its paucity of machinery development. A second UK company, Wilson and Longbottom, by contrast, pursued a vigorous innovation policy, and is now technical and market leader in several specialist loom areas.'

Pavitt *et al.* (1980) have provided further documented evidence of the linkage between technical innovation and economic performance. Several studies showed convincingly that overseas competitors had grown stronger than British companies in international markets largely because of their greater emphasis on a policy of vigorous product development. The authors called for greater attention to improving technical innovation in the United Kingdom. In Pavitt's view this would require 'a fundamental transformation of British attitudes, skills and resources committed to technical innovation'.

Management's approach to technical innovation

The study of technical innovation has increased rapidly in recent years to the extent that it is now possible to identify three alternative (mutually complementary) reasons for the apparent inability of British companies in some sectors to develop new products of improved quality and technical sophistication.

The first possibility is that British companies in these sectors do not manage the new product development process as well as their international competitors. In this interpretation, blame is laid at the door of the supplier. This view has been a popular one amongst commentators who have argued that British industry's approach to new product development, in particular in monitoring technical change and changes in market requirements, is less competent than that adopted by overseas competitors.

An alternative view suggests that one of the major problems in developing and marketing technological innovation is the attitude of potential users to technical change. No matter how attractive the new product may appear to its manufacturer on technical and economic grounds, the potential customers' attitude to the innovation will influence how rapidly the product will be accepted.

If, as the evidence cited below suggests, the supplier relies on domestic customers to help with ideas for new products, and solicits their opinions on new developments as they proceed from drawing board to prototype to final product, then a conservative attitude on the part of domestic customers will inhibit the rapid development of new technology. One of the barriers to developing new, higher quality products in several sectors in the United Kingdom may therefore be the inherent conservatism of the British customers.

A final possibility is that since successful new product development and adoption decisions appear to depend on a high degree of interaction between supplier and user, it is the quality of that interaction in itself which determines whether or not the right product quality is developed. It could be argued that British companies are not as successful in new product development as their overseas competitors because the quality and extent of interaction between themselves and their domestic customers is lower. In other words, successful new product development is a joint responsibility of both buyer and seller.

Outline of the book

Successive chapters of this book develop each of these three themes in succession in a comparative review of supplier and customer involvement in new product development decisions in the United Kingdom. Such a comparison is intended to highlight major differences between the approach adopted in British companies and that of one of the United Kingdom's most successful overseas competitors. Such differences may then provide an indication of possible remedies in the United Kingdom as well as further insight into the necessary conditions for successful product innovation in general.

Chapter 2 briefly reviews the literature on new product development and purchasing decisions in industrial markets to establish the framework for this research enquiry. Chapter 3 describes how the field research was undertaken, beginning with an elaboration of the basic problems of research into new product development and purchasing decisions. A combination of methods was used, and this chapter details which techniques were chosen and the problems experienced in implementing them. Chapter 3 also discusses why a decision was taken to focus on only one industry in the subsequent fieldwork and justifies the choice of the machine tool industries of the United Kingdom and West Germany.

In Chapter 4 the results of a survey of suppliers are analysed. Comparisons are drawn between the approach of British and West German machine tool manufacturers to new product development. These comparisons suggest significant differences between the two industries which are related to success in new product development.

Chapter 5 focusses on a survey of customers in the United Kingdom and West Germany. The results of this study support the view that one of the problems of the British machine tool industry may be the inherent weaknesses of its own domestic customers. Chapter 6 examines the final proposition which suggests that the basic problem in the United Kingdom may be a lack of integration and interaction between customers and suppliers. Finally, Chapter 7 examines the implications of the study for the machine tool industry and for other sectors of manufacturing industry. Government policy in encouraging technological innovation is also re-examined in the light of these findings.

2

New product development and purchasing decisions

Introduction

In the previous chapter it was suggested that there were three basic interpretations of the comparative weakness of British companies in developing new products to meet the requirements of the market. Each of these three interpretations is based upon a different research tradition, and the purpose of this chapter is to review what is currently known about each of these areas as a preliminary to the fieldwork reported in subsequent chapters. This review is intended to guide the reader to the current knowledge of factors related to success in new product development, factors influencing the relative rate of acceptance of technical innovation by customers, and finally factors influencing the extent of interaction between supplier and customer.

Supplier interpretations

The new product development process is generally conceived as a series of sequential steps from idea generation and initial screening, through business analysis to prototype development and test marketing, and finally commercialisation. The process itself may approximate to a more rudimentary one than this in many firms, but each of these activities could be expected to be present to a greater or lesser extent. Given this range of activities, it is clear that different functions from within the organisation will become involved in each activity. There will therefore be a requirement to integrate the efforts of different departments effectively.

Organisations can vary in the techniques used to provide the necessary integration. Alternative formal organisations include 'venture' management, new product committees and matrix management. In the venture management approach, in effect a 'mini business' is created in the firm with its own staff and budget to manage the product line. The venture group will usually contain representatives from each management function in the firm. The new product development committee brings together specialists from different functions on a regular basis to plan new products and control their developments, but managers have other responsibilities as well as product development. The matrix management approach is an organisational form chosen to give managers from different functions responsibility for different aspects of the same problem.

However, it is possible that many organisations may have no formal separate organisation for product development decisions, perhaps taking them on an *ad hoc* basis in general management committees or on the responsibility and initiative of individual managers.

The relationship between organisational structure and innovativeness has been extensively studied in the administrative science literature and more recently in the marketing literature. Broadly speaking, it would appear that there are certain characteristics of structure which influence an organisation's receptivity to innovation.

Burns and Stalker's (1961) categories of 'organic' and 'mechanistic' organisations have been widely used to predict how receptive a company would be to the introduction of an innovation. In their model, the 'organic' type of organisation has a greater degree of informality and shared responsibilities than the 'mechanistic' one. It has fewer hierarchical levels than the 'mechanistic' organisation and information flows freely. As a result of these characteristics, it has been widely suggested that an organic organisational structure may make a company more receptive to change and more responsive than a mechanistic structure. Conversely, once the organisation has generated the idea, a more mechanistic structure appears to be required to translate the concept into a viable product. Johne (1982) has identified the importance of companies being able to make this mechanistic shift in a study of product development in the electronics industry.

Whilst organisational structure may have a marked impact on the innovative performance of companies, there is considerable evidence to suggest that other factors may be equally important. These factors include the attitudes and general 'quality' of management itself. Carter and Williams (1957) established twenty-four factors which were present in technically progressive firms and not present in unprogressive firms.

These factors included high quality incoming information, a deliberate search for new ideas, a willingness to share knowledge, a readiness to look outside the firm, effective internal communication and coordination, a high status given to science and technology in the firm, and a consciousness of costs and profit in the research and development departments (where these existed). Other factors included a sound policy for recruitment of management, an ability to attract people and a willingness to arrange for the effective training of staff.

Organisations with these characteristics, according to Carter and Williams, tended on average to be more innovative and receptive to change than those without. Subsequent research has tended to confirm this view. General management quality appears to influence innovativeness strongly as the two authors commented: 'To a great extent the receptiveness of a firm to technical change must depend on the understanding and appreciation of science and technology to be found in the various ranks of management, and this in turn depends on the training and experience of individual managers.'

In terms of the techniques of new product development the continuing high failure rate of new products justifies the assertion that there is as yet no one prescriptive approach to the management of the new product development process. There is, however, a large body of literature addressed to separate aspects of this process. A review by Robertson (1977) of case-study-based research on innovation synthesises much of the earlier work in this area.

From this review it is clear that successful innovation is more likely when close attention is paid to user needs at all stages in the development process. Success also appears to be more likely when technical knowledge and research and development are high status activities in the company and top management support exists for new developments. Organisations which deliberately scan the environment looking for new ideas are also more likely to be successful. Finally, and crucially in many observers' opinions, success is more likely where a 'product champion' exists to support the product through various stages in its development. In short, innovation appears to be more likely to be successful where there is good communication between managers, good understanding of user needs, and good quality management.

More recent work by Cooper (1980) identified three composite factors which appeared to dominate the 'success equation' for new industrial products. These were product, market and technical strengths. The product dimension referred to the extent to which the product was perceived as significantly different from competing products and was measured by the product's comparative advantages over existing conventional solutions. The market dimension referred to the extent to which adequate market analysis and customer contact had been undertaken during the various stages of product development, and the technical dimension referred to the technical skills of the supplier and the extent to which such technical skills were brought to bear on the development of the new product.

There is now a considerable literature extant on the different techniques of product development. Writers such as Twiss (1974), Pessemier (1966), and Baker and McTavish (1976) have presented detailed analyses of the process. The author's synthesis of this literature is presented in Table 2.1. The items in this table will be used in a later chapter of this book as a yardstick with which to compare the product development activities of the British and West German companies in the sample.

Despite the existence of this large body of literature on systems and models to be used in new product development decisions, there is evidence to suggest that managers may often disregard such techniques in their actual decision taking. For example, Robertson and Fox (1979) commented:

> No such techniques were applied by the managers we observed, and in the 38 projects examined in detail during the Sappho project (1971) there was a marked absence of evidence that such techniques had been used. To this extent it can be concluded that formal decision making techniques are not among the systems used by managers in arriving at the decisions which initiate and guide industrial innovations.

The general consensus on this aspect of the innovation process appears to be that there is as yet no one set of techniques that could be used to reduce the uncertainty of new product development.

Table 2.1. *Principal techniques appropriate to new product development process (a summary of techniques suggested by the literature)*

1. Idea generation

a. Consider product planning as part of corporate strategy.
   - Are there 'gaps' perceived in company's present or future offerings?

b. Systematic review of technology.
   - Monitor scientific literature, attend conferences, seminars, participate in trade associations, sponsor university research.
   - Forecast new technologies using forecasting techniques, e.g. Delphi, trend extrapolation, morphological analyses.
   - Set R and D budgets.
   - Run creativity sessions to 'create' ideas – or 'buzz' sessions.
   - Monitor changes in supply markets.
   - Review patents and new licence possibilities.
   - Monitor competition.

c. Analyse market.
   - Systematic 'lost order analysis' of sales trip reports.
   - Continuous data collection from identified market segments.
   - Special *ad hoc* surveys of consumer groups/individual consumers.
   - Run creativity sessions to 'create' new ideas.

Much appears to depend on the quality of individual managers involved in the development work. As Carter and Williams commented some 20 years previously (1957): 'Where the uncertainty is not so easily dismissed, the decision to commit resources, as we have seen it in the case studies, depends on a number of personal and general factors. The best way to ensure that a project with great uncertainty attaching to it is carried through is to have a final decision maker who believes in it.'

Research issues

The preceding sections have highlighted some of the issues raised by an extensive review of the literature reported elsewhere (Parkinson, 1980). This review suggested that the comparison between British and West German suppliers in the field research should concentrate on three related issues. These were the systems used to organise and control product development, organisational structure in a general sense, and general management quality including the

Table 2.1 (*cont.*)

2. Initial screening
Measure consistency of product idea with overall objectives of company.
Produce summary product appraisals which:
1. estimate profitability
2. estimate sales performance
3. estimate capital requirements
4. estimate demands on resources
5. evaluate technical opportunity
6. evaluate patent protection
7. estimate production load
8. evaluate impact on existing products and similarity of major business

3. Business analysis
Use various techniques to estimate:
1. development costs
2. manufacturing costs (including prime costs and additional capital)
3. market price, sales volume, break-even point, duration of sales
4. sales and launch costs
5. cash flows out of and into company
6. use sensitivity analysis to compare effect of different assumptions on 1-5

4. Product development
Ensure value for money in design by:
1. value engineering of product
2. continuing tests of consumer response as product develops
3. prototype evaluation and testing

5. Market launch
Pre-identify target market/test market.

attention paid to the use of formal techniques in new product development.

Buyer interpretations

Historically, there have been two distinct approaches to the analysis of decisions by companies to adopt new products. The first of these has focussed on economic and technical determinants of choice, and sought to explain buying decisions in terms of the relative economic and technical merit of new investment proposals. The second approach has emphasised the effect of other essentially 'managerial' factors on adoption decisions, and has shown how management's perception of the economic and technical significance of innovation may be a more relevant factor in explaining choice than economic or technical considerations in an absolute sense.

There is considerable evidence in the literature to support the view that economic and technical features play an important role in influencing the speed with which an organisation decides to adopt an innovation. As Table B.1 in Appendix B shows, researchers have found significant positive relationships between the rate of purchase of an innovation by individual firms and a range of economic factors.

For example, Mansfield (1963) found significant relationships between the profitability of the firm, the size of the investment, the size of the firm, and the expected return from the investment and the rate at which organisations in several sectors had adopted specific innovations. The innovations selected for study by Mansfield were selected on the basis of recommendations from 'experts' in the industry who were asked to cite examples of 'significant' innovation. Therefore, the innovations which were selected were not necessarily important or significant to the potential user.

Additionally the statistics on each firm were collected by the researcher on the basis of an external review of the information available. Thus it was assumed that the absolute level of profitability or rate of sales growth of the company reported in published accounts would be a useful predictor of the rate of adoption. The researcher also made his own estimates of the profitability of the innovation to each company, based on his own view of the position of the company. With the benefit of hindsight it is easy to see, as Mansfield himself has commented, that this approach may be misleading and that it is management's perception of these factors rather than their absolute level which is crucial. In other words management's attitude to the specific innovation has an important mediating effect. Thus, for example, a new product or process offering 20% return on investment to a company currently earning 15% will tend to be perceived differently in that company from the way it would be in one where management was used to a 25% return.

Since Mansfield, other researchers have increasingly incorporated the potential user's response to innovation in their analysis rather than their own perception of the significance of the innovation. Abu-Ismail (1976), for example, found a significant relationship between the user's perception of the potential benefits of an innovation in the textile industry and the rate with which individual organisations adopted the innovation (a new type of tufted-carpet backing material). The perceived benefit was an improvement in quality in the end product which would be appreciated by the companies' own customers.

More recently the author's own work with Professor M.J. Baker has confirmed the relative importance of the users' perception of the economic and technical benefits of innovation as influences on the rate of adoption of industrial innovation. Studies of innovations in the textile industry, the building and construction industry, metal-forming, and latterly in the offshore oil industry have all shown that organisations where managers could see a high level of potential economic and/or technical advantage in adoption of an innovation would be amongst the first to adopt (Parkinson, 1982).

It is also appropriate to examine the results of a further study which shows the relative importance of a range of general factors in the investment decision. The author participated in a broadly based study of investment decisions in British manufacturing industry in which respondents were asked to indicate the relative importance of a series of technical and economic factors for any one specified investment decision (Baker and Parkinson, 1976). The overall response to this survey is detailed in Table B.2 in Appendix B.

From this survey it was clear that the most important factors considered in investment decisions by a sample of over seven hundred companies in British manufacturing industry were 'reliability', 'quality of output', and 'return on investment'. Technical criteria were ranked ahead of economic or commercial criteria in investment evaluation, suggesting that for the sample as a whole product quality and performance were more important than other essentially financial criteria. Thus a general survey of investment decisions provided more support for the general theme of the investigation reported in subsequent pages, supporting the need for firms to concentrate more on product quality than on price competition *per se*.

Whilst economic and technical influences are clearly important in industrial adoption decisions, the economic and technical advantages of an innovation over conventional alternatives do not provide a consistent explanation of the rate of acceptance of innovation.

Mansfield, for example, could only explain 50% in the variance of the rate of acceptance of the innovation he studied by economic variables and was led to conclude that other 'managerial' factors may be important (1963). He commented:

> Perhaps these variables (profitability, liquidity and growth rate) are less important than other more elusive and essentially non-economic variables. The personal attributes, training and other characteristics of top and middle management may play a very important role in determining how quickly a firm introduces an innovation. The presence or absence of a few men in the right place, who believe strongly in the value of a new technique may make a crucial difference.

Since this original observation, a considerable amount of work has been done in an attempt to develop a better understanding of the ways in which management's attitude may influence the rate of acceptance of innovation. In this area, many of the findings from the innovation (new product development) literature are readily transferable.

Table 2.2 summarises some of the principal research findings which relate organisational characteristics and the rate of adoption of industrial innovation from outside the firm. The studies reported in this table have all shown that 'managerial' factors can play an important role in influencing the speed with which an organisation may adopt an innovation. Each of them has also combined earlier 'economic/technical' factors with 'managerial' ones in an attempt to produce a useful predictive model – with varying degrees of success.

Research issues

As with the previous discussion of the supplier dimension of this problem, this review is a summary of an extensive literature search and previous work in this area by the author. The purpose of the review was to highlight the basic dimensions of organisational purchasing behaviour which would be examined in the comparison of British and West German customers. In the comparison between the two countries emphasis will be placed on several factors. These include the relative importance given to technical and economic considerations in customers' evaluation and choice of capital equipment. They also include organisational characteristics such as structure and climate. Finally, attention will also be focussed upon managerial characteristics such as the use of formal procedures to control the actual purchase of new equipment, and the collection and use of information in new machine purchasing decisions.

Interaction interpretation

The 'supplier interpretation' and 'user interpretation' described in the preceding pages represent the two current interpretations of the reasons for failure in new product development. In short, either the supply industry is inadequate on some dimension or the customers themselves are resistant to new technology in the domestic market and so inhibit its acceptance.

However, there is a third viewpoint, based on treating the customer and supplier as engaged in a joint activity in product development. Instead of treating customer and supplier activities in isolation from each other, it may be more useful to regard them as an integrated activity. Specifically, the quality of integration between customer and supplier could be used to predict the success or failure of new product development and adoption decisions.

Von Hippel (1977) develops this theme in a discussion of the extent of user involvement in process equipment innovation. According to von Hippel the user may play a considerable role in the development of new technology through perceiving the need for the product innovation, and conceiving a solution. Proactive customers may go to the extent of building a prototype and proving the value of the prototype by using it. In using the prototype the customer may also diffuse detailed information on the value of the development and on how it can be replicated to other potential users and to firms which might be interested in manufacturing the development on a commercial basis.

In most situations the customer is unlikely to go to the extent suggested here without some supplier involvement in engineering work on the customer's design, to improve reliability or reduce costs. The extent of involvement of the supplier may be considerable once the original idea has been suggested by the customer, particularly in high technology industries where the supplier's specialist knowledge is likely to be far greater than that of the customer. However, von Hippel's discussion adds an important dimension to the

Table 2.2. *Studies of the relationship between organisational characteristics and the purchase of industrial innovation*

	Organisational characteristics	Hypothesised relationship to receptivity to new products
Baker (1975)	Firm's attitude to generation acquisition and evaluation of information	Positive
	Organisational structure (lower degree of formality leads to greater extent of early adoption)	Positive
	Prior education and training of employees	Positive
	Willingness of company to publicise own research findings, and participation in joint ventures	Positive
	Emphasis placed on use of scientific management practices, formal planning systems, purchasing as a business function, quantification of investment decisions	Positive
Parkinson (1975)	Quality and quantity of information available to decision takers	Positive
Abu-Ismail (1976)	Degree of centralisation	Negative
	Degree of formalisation (use of written procedure)	Negative
	Use of good communications system within the firm	Positive
	Degree of integration between departments involved in the innovation process	Positive
	Orientation of company	
	– marketing	Positive
	– technical	Positive
	– production	Negative
	Personnel policies advanced	Positive

Table 2.2 (*cont.*)

	Organisational characteristics	Hypothesised relationship to receptivity to new products
Baker and Parkinson (1976)	Organisational structure – degree of structure	Negative
	Use of formal techniques	Positive
	High quality of information available to decision takers	Positive
	Organisational formality (use of formal reviews)	Positive
	Attention paid to research and development priorities	Positive
	External orientation of executives	Positive
	Attention paid to marketing	Positive
El Sherbeny (1978)	All the factors cited in Baker, and Baker and Parkinson in this table, and additionally creativity and the attitude of top management	Positive
Gronhaug (1976)	Principally organisational structure	Unspecified

examination of the process of product development. The supplier and user interpretations tend to view this process in terms of sellers or buyers in isolation. This third interpretation brings seller and buyer activities together in a focus on joint activities.

Relationships between a firm and its customers evolve over time and usually exist independently of individual transactions between buyer and seller – although clearly each transaction has an important impact on that relationship. Such relationships are a form of inter-organisational behaviour – an area which has come under increasing study in the administrative science literature in recent years. Findings from this literature are also appropriate to the study of product development decisions.

In particular, the work of Aiken and Hage (1968), and Pugh *et al.* (1968, 1969) provide useful insights into the relationship between

Table 2.3. *Major findings of Aiken and Hage and interpretation in innovation/adoption context*

Finding No. 1	Organisations with many joint programmes are more complex organisations, that is they are more highly professionalised, and have more diversified occupational structures.
Interpretation	In industrial markets those organisations which work together to solve a common problem (for example, to develop a new product) will tend to be more highly professionalised and have more diversified occupational structures (in short will give higher status to individual professional qualifications, and have a greater range of talents and skills available), when compared to organisations with fewer joint programmes.
Finding No. 2	Organisations with many joint programmes are more innovative organisations.
Interpretation	As the degree of interaction between buyer and seller increases so the probability that some of this communication will lead to ideas for new product development increases.
Finding No. 3	Organisations with many joint programmes have more active internal communications channels.
Interpretation	Buyer–seller interaction will be highest between those organisations which themselves have effective and active channels of internal communication.
Finding No. 4	Organisations with many joint programmes have slightly more decentralised decision-making structures.
Finding No. 5	Organisations with many joint programmes have slightly less formal decision-making structures.
Interpretation of 4 and 5	Buying or selling organisations which have essentially organic characteristics are more likely to engage in joint programmes of new product development.

organisational structure, organisation interdependence and new product development decisions. Aiken and Hage's findings and their interpretation in the context of new product development appear in Table 2.3. In this table, Aiken and Hage's findings from their study of general interorganisational behaviour are interpreted

Table 2.4. *Major findings of Pugh et al. and interpretation in innovation/adoption context*

Finding No. 1	Organisational interdependence is negatively related to the size of the organisations between which joint programmes are feasible.
Interpretation	Buyer and seller are more likely to be interdependent if both (or either) are relatively small companies.
Finding No. 2	Organisation interdependence is negatively related to the degree of centralisation of the organisations between which joint programmes are feasible.
Interpretation	Buyer and seller are most likely to be interdependent if both (or either) are relatively decentralised companies.
Finding No. 3	Organisational interdependence is positively related to the extent to which the organisations are dependent on new technology.
Interpretation	Buyer(s) and/or seller(s) which are dependent on new technologies for their continued survival are more likely to be interdependent.

in terms of interorganisational behaviour between supplier and buyer in new product development.

The work of Pugh *et al.* on organisational interdependence and its interpretation in the context of new product development is summarised in Table 2.4. Organisational interdependence, amongst other factors, measures the extent to which organisations rely on help from each other in a changing environment. As the pace of change increases, then, all other factors being constant, so organisational interdependence increases.

Pugh *et al.*'s findings indicate that organisational interdependence (in this context, joint programmes between supplier and buyer of new technology) will be greatest when both companies are relatively small, decentralised organisations. Interdependence will also be greater when both are dependent on new technologies for continued survival.

Relatively recently (since 1974), research in the area of Marketing has begun to acknowledge the relationship between buyer–seller interaction and new product development. According to Mattson (1975) an organisation's decision to adopt new technology will depend not only on the relative attractiveness of the new technology to the organisation, and the receptivity of the organisation in a general sense, but also on the stability of the relationship between buyer and seller. In Mattson's model new product development and purchasing processes cannot be separated, since the potential customer plays a major role in the development of the new product. As a relationship between buyer becomes more stable, so the uncertainties of new product development and purchasing decisions appear to reduce, enhancing the rapid development and acceptance of new technology.

Hakansson and Ostberg (1975) have summarised this process thus:

> The perceived uncertainty in a situation which is dependent on a given set of variables may be reduced by a social exchange process and/or by adaptations. Purely economic considerations may further motivate adaptations. Both the processes of social exchange and the engagement in adaptations will then often lead to very solid commitments between buyer and seller, that is they tend to build up extensive and lasting power dependence relationships.

Finally, and most recently, the concept of exchange between buyer and seller in general selling and buying situations has been investigated in a broadly based study of marketing and purchasing decisions in several different European countries (Hakansson, 1982). This work is one of the first to take a structured sample of companies in different European countries and compare the relationships which exist between buyers and sellers in each country and between countries. In this book the authors develop an interaction model which identifies the importance of several factors including the environment within which companies operate, the types of companies involved and the relationships which exist between these companies. The structure and manufacturing technologies of individual companies are also seen as having a major influence upon the nature of interaction between companies. The model provides a useful framework within which to analyse the interaction between buyer and seller, and further discussion of this model appears in the analysis of buyer–seller relationships in new product development in Chapter 6.

Research issues

The interaction interpretation of new product development focusses attention on the relationship between buyer and seller, rather than on buyer or seller in isolation. The literature raises two basic issues which will be examined in the field research in Chapter 6. These are the extent and type of interaction which exists between companies and their customers and the factors which appear to influence that interaction. The implications of preceding research are that attention should be paid to the degree of concentration in the numbers of buying or selling companies, the relative size of buyer and seller, and organisational and managerial characteristics of the type already discussed in the supplier and customer interpretations.

3

Research design

Alternative research methods

At the outset of this study it was evident that there were a considerable number of options for the research design. The one which was finally chosen represented a compromise between the need to collect information which would be useful in a broadly based comparison of British companies with their overseas competitors, and the cost and resources available. Figure 3.1 summarises these alternatives.

The ideal design assuming no resource constraints would have covered several key industries rather than one, focussed on specific products in each industry, and compared user–supplier relationships in more than one overseas country with those relationships in the United Kingdom.

The advantage of studying more than one industry clearly lies in the broader base of information from which generalisations can be drawn. Comparisons between different industries could also be expected to yield more insight into the influence of variables in each industry which may make the processes of new product development and introduction unique to the industry. In short, it would

Figure 3.1. Alternative research designs

Design factor 1	Multi-industry study or individual industry study
Design factor 2	Product-specific study or non-product-specific
Design factor 3	Multi-country comparisons or single country comparisons

allow greater examination of situational variables such as the degree of supplier and/or customer concentration.

Product-specific studies would focus on the development and acceptance of a specific innovation(s) introduced, or on the point of being introduced, to the market. Once again, given that it is apparent that the final decision to develop or adopt a new product ultimately appears to depend on situational factors which are unique to the new product, focus on specific products or processes would help to identify a range of factors associated with those products or processes, as well as providing a focus for more broadly based generalisations about the process as a whole.

Comparisons between the United Kingdom and more than one overseas competitor would again allow detailed consideration of variables in different countries which may make the processes of new product development and diffusion unique to the country. Cultural variables, for instance, could be studied in more detail than would be possible in a simple two-country comparison.

Therefore, there would have been many potential attractions in a multi-country, product(s)-specific, multi-industry approach to the problem. However, because this was still an essentially exploratory investigation, it was felt that the study should be restricted to comparisons between the United Kingdom and one other country. The enquiry would not be focussed on the development or adoption of the same specified products, although it would involve asking suppliers and their customers to use specific examples to illustrate their own decisions in product development or introduction situations. Finally, the study would be limited to one industry only. This research was designed therefore to provide the basis for future investigation of buyer–seller relationships in new product development on a wider basis.

Despite these obvious limitations, the basic design of the study was still significantly different from earlier studies. There are, as yet, few reported studies of buyer–seller interaction in new product development decisions. There are even fewer studies which have adopted an international comparative approach.

Longitudinal and retrospective studies

Two basic alternatives were considered for the actual fieldwork. The data could have been collected retrospectively by visiting companies and/or administering a postal questionnaire to ask managers about their perception of the events associated with the development or adoption of an innovation. This was the approach adopted by the majority of previous research in this area. Such data is deficient to the extent that some degree of *post factum* rationalisation will inevitably take place, distorting the respondent's perception and reporting of the events under examination. A second problem is that if events took place over a period of years, the personnel involved with specific decisions could have moved on and be difficult to trace for interview.

Alternatively, the data could have been collected longitudinally as events took place. The processes of new product development and introduction would have been observed over time. This would avoid problems of respondent recall, but there are several fundamental drawbacks to the use of this approach. The researcher hopes that during the time period that the longitudinal study is being conducted, product development or introduction decisions take place. Therefore, broad coverage of suppliers and their potential customers is required, since it would be very difficult *a priori* to identify where such decisions might take place. There is also the problem of the time needed to set up and administer a longitudinal study, both in securing the initial cooperation of managers and in making the necessary observations. The international focus of this study would have exacerbated this problem.

A longitudinal approach requires a considerable degree of commitment from cooperating managers. Since the researcher cannot physically be present at all events he is interested in recording (some may take place at the same time in different places), some form of diary is usually used to collect information. Managers in cooperating firms are issued with a diary and asked to complete it regularly, recording salient events as they occur. Information collected from the diaries is usually supplemented by more detailed questionnaire-based data collected at regular intervals by the researcher.

Despite the attractions of a longitudinal study, the basic characteristics of this approach were unsuitable for this particular research enquiry. A decision was taken to adopt a retrospective form developing case studies of new product development processes in supply companies, and administering a postal questionnaire to collect data on purchasing decisions in customer organisations. As subsequent comments indicate, the supply industries were both relatively concentrated with a large proportion of the output of the industry coming from relatively few companies, favouring an approach based on individual visits to each of the leading companies in the industry. On the other hand, the products of the industry were used in a wide cross-section of manufacturing industry by many different types of customers, and therefore a postal questionnaire approach was felt to be more appropriate to reach a broad sample in each country (although case studies of the purchasing decision were initially prepared from visits to selected customer organisations, and these case studies were used to develop the postal questionnaire).

Choice of industry

Three main factors influenced the choice of industry to be studied. The industry had to be important to the British economy as a whole in terms of its employment and contribution to the National Product. It also had to show a pattern of decline in terms of a reducing share of the international market for its products, and, associated with this, a history of gradually reducing product quality, in terms of being increasingly less able to supply products suited to customer needs than its overseas competitors.

The machine tool industry was chosen for this study since it met each of these three criteria. The industry itself has a strategically important position in the national economy, since its products are used by virtually the whole of manufacturing industry. A strong and progressive domestic industry is clearly, therefore, important to the well-being of the whole economy. Additionally, the British machine tool industry has contracted dramatically in size in post-war years. Employment fell from 87,000 in 1960 to 44,000 in 1982, and the British share of the world export market has also contracted considerably (for example, falling from 8% in 1971 to 3% in 1981).

The technical sophistication of the product measured in terms of value per ton has also fallen significantly compared with products from overseas competition. For example, in 1977 the value of exports from the United Kingdom averaged £3500 while imported machines averaged £4100. More significantly perhaps, one of the main reasons given by domestic buyers for buying from abroad rather than buying from British companies in a survey for NEDO

(1977) was the relatively higher technical sophistication of the overseas product.

Choice of overseas competitor

The country chosen for comparative purposes in the field study was West Germany. The Japanese machine tool industry would also have been a suitable choice, but language barriers were an obvious problem. In contrast to the position of the British machine tool industry, the West German industry has become one of the strongest in the world, largely owing to a continuing policy of technical sophistication in its products which has created a reputation for 'German quality'. Despite the strength of the Deutschmark against other currencies, the industry has consistently exported more than 35% of its output.

Design of the supplier survey

Both British and West German machine tool industries are relatively concentrated, with comparatively few firms (or major groups of firms) accounting for a significant proportion of the output of the industry. This pattern of distribution indicated that a case study approach was the best form of research to collect the data required. Visits were made to the largest companies in each country, and case studies were prepared detailing each company's approach to product development.

Data collection was restricted to the largest companies for two reasons. Firstly, it was felt that new product development was likely to be at its most formal and 'professional' in the largest companies in the industry. Size was felt to be a surrogate indicator of the extent of use of professional management techniques and sophistication – an assertion supported by the work of previous researchers. Secondly, since these firms contributed most to the output of the industry, a comparison between new product development activities in each company would provide useful guidance to policy.

In summary, the nine companies in the West German sample accounted for approximately 35% of the employment of the West German machine tool industry in 1977, and the seven companies in the British sample were members of groups which between them accounted for 58% of the output of the British machine tool industry in 1978.

The case study material which was developed from the field research represented a distillation of the opinions of individual managers in each company visited, together with other information collected from a wide variety of secondary sources of information, including company reports and accounts, the trade press, key government reports, and other permanent documentation made available to the researcher by individual companies. A semi-structured questionnaire was used to guide the discussion, and therefore there are some key standard questions which were asked in each company. The number of managers interviewed in each company varied from two to six, depending on how many were involved in new product development, and on their availability.

Each of the case studies followed the same general format, beginning with general characteristics of the supplier organisation, discussing its history, patterns of organisation, communication and control, as antecedents to the product development process. The case studies then examined the relative contributions of Marketing and Sales, and Research and Development to new product development. The new product development techniques and procedures of each company were detailed, identifying the participation of individual managers at different stages in the process, and the criteria seen as important in new product development. In particular, attention was given to the relative importance accorded to individual product attributes. Finally, the case studies discussed the extent to which each company saw its customers as being involved in product development.

Design of the customer survey

The wide distribution of potential machine tool customers meant that a case-study-based approach was impractical within the time and resource constraints of the study. It was also expected that the response to a postal questionnaire would be low. However, even with a relatively low response rate it was felt that a postal questionnaire approach could generate some useful insights into differences between West German and British customers which would not have been possible from a more limited series of case studies.

The two samples were drawn at random from suitable sampling

frames in each country. The British sample was weighted by the distribution of machine tools in each industrial sector. Comparable figures were not available to stratify the German sample. In both countries a single copy of the questionnaire was sent to the managing director or general manager of the company, with a request that the questionnaire be passed to the person(s) most concerned with the purchase of new machine tools. Clearly this technique did not allow more than one manager to reply to the survey, and in most situations more than one manager would have been involved in the decision process.

However, there are fundamental problems in aggregating the response of more than one individual. For example, average scores do not take into account differences in status or participation amongst respondents. Consequently it was felt more appropriate to ask the individual to detail the extent to which other individuals were involved in the purchasing decision and the nature of their involvement. In this way the survey did produce a summary picture of the decision-making unit in the firm. Moreover a considerable number of questions were specific and could be answered by one respondent.

Once the questionnaire had been piloted and reworked where necessary, it was translated into German and retranslated into English to check for misinterpretations in translation.

Reflecting the nature of the enquiry, the postal questionnaire contained questions which were in effect the parallel or reverse of those covered in the supplier study. The first questions covered organisational structure, patterns of communication and control, this time as antecedents to the machine tool purchasing process. The next section asked questions about the relative contribution of different departments to the purchasing decision and this was followed by a series of questions on the purchasing process itself. Customer perceptions of the relative importance of different product attributes were also measured, and finally several questions were asked about customer involvement in the suppliers' new product development decisions.

The overall response to the survey is detailed in Table 3.1. From this table it can be seen that this was relatively low, as had been expected, improving from a 10% usable response at the first mail out to a 16% usable response in the follow-up (73 British companies and 56 West German companies). However, the number of

Table 3.1. *Response to postal survey of customers*

	First mail out		Follow-up		
	United Kingdom	West Germany	United Kingdom	West Germany	Total
Mail out	400	400	324	348	
Usable replies	47	36	26	20	129
Gone away	5	4	2	—	11
Other replies:					
Do not buy machine tools	24	12	58	29	123
Other reasons	—	—	20	18	38
Total	76	52	106	67	301

companies in the sample was far greater than could have been produced using a case study approach, and the two samples are broadly comparable as the analysis in Chapter 5 shows.

The main reason given by respondents for not completing the questionnaire was that their company did not buy machine tools. Consequently, the sampling frames were deficient to the extent that they included these companies in the sample, and the actual response from machine tool buyers was higher than it appeared from the overall statistics.

Despite these limitations, the two samples do cover a wide spread of industries using machine tools, as Table 3.2 indicates, and there is some comparability between the samples (although the West German sample had proportionately more respondents from the machine tool industry, and the motor industry, whilst the British sample had more respondents from Industrial Plant and Steel Work, and Metal Industries not elsewhere specified).

In terms of employment, analyses of the size of the companies in the samples against the known size distribution of companies in the industries from which they were taken, indicated that both samples tended to over-represent larger organisations.

In the interpretation of results which follows, therefore, it is clear that *no claim can be made for a fully representative sample*. However, it is believed that the breadth of data for each company, in terms of the number of issues dealt with in the questionnaire, and

Table 3.2. *Analysis of response by industrial sector (%)*

	West Germany	United Kingdom
Agricultural machinery	0	0
Metal-working machine tools	16	6
Industrial engines	2	3
Textile machinery	4	1
Construction equipment	2	4
Mechanical handling	2	6
Office machinery	2	0
Industrial plant steelwork	7	20
Other mechanical engineering	13	20
Scientific and surgical instruments	2	3
Electrical machinery	11	7
Telegraph and telephone	0	1
Domestic electrical appliances	2	1
Other electrical	4	3
Shipbuilding	0	1
Motor vehicles	25	4
Aerospace	0	1
Locomotives	2	0
Small tools and gauges	2	3
Hand tools	0	1
Bolts/nuts/screws	0	1
Wire/wire manufacture	0	1
Cars and metal boxes	0	0
Metal industries (not elsewhere specified)	4	13
Total	100 (56)	100 (73)

the number of firms in each sample, does allow generalisations to be made about the differences between British and West German machine tool customers in their attitudes to technological innovation.

4

Patterns of technological innovation in British and West German machine tool manufacturing

Introduction

Chapter 4 examines the differences between British and West German machine tool manufacturers in their approach to new product development indicated by this study. The chapter begins with a discussion of environmental influences on product development, focussing on the structure of each industry, patterns of management education, and the role of trade associations, universities and other research institutions. Attention is then focussed on individual companies, and comparisons are made between patterns of ownership, export trading performance, product range, degree of customer specialisation and organisation. Each of these aspects of the companies studied forms an important part of the context within which innovation decisions are taken. Focussing the discussion still more precisely, attention then turns to differences in the new product development process itself.

Environmental differences

The patterns of concentration in the British and West German machine tool industries differ considerably. In both countries, relatively few companies account for the largest proportion of the output of the industry, but in the United Kingdom these major companies are often groupings of smaller machine tool manufacturers with varying degrees of independence from group policy. By contrast, in West Germany there are fewer conglomerates, and a large proportion of the output of the industry comes from medium-sized independent manufacturers employing between 501 and 1000 people.

In the United Kingdom successive reports on the machine tool industry have emphasised that concentration was necessary to achieve the economies of scale needed to remain competitive. For example, a section of the 'Way' report on the industry (1970) suggested that:

> Scope exists for product rationalisation and concentration because:
> (a) The cost of developing, manufacturing, and selling advanced machines and systems of production is such that a high market share is needed to support the effort needed.
> (b) The competition in the conventional general purpose machine tool field is such that large scale production is necessary if the British industry is to remain a viable supplier at home and abroad.

However, the rationalisation which has taken place in the British machine tool industry appears to have produced a barrier to technological innovation rather than stimulating it. In the discussion of the management of product development which follows British machine tool manufacturers, although part of a group, do not pool technological or managerial resources in the way suggested by the 'Way' report. Rather they continue with individual development work at local sites, often with a low level of resources.

The structure of the West German machine tool industry appears to be more conducive to technological innovation, as K. Gooding (1978) commented:

> Couple this [accounting practices] with the immense flexibility of the German industry, made up as it is in the main of medium sized family companies able to make swift decisions, and act on them immediately, and it is not difficult to understand why some of its European neighbours have seen West Germany as the major threat to their home manufacture.

The individual case studies of the West German companies in this study support this viewpoint. The companies in the sample were more likely to be independent public companies. In addition there

Table 4.1. *Ownership category of firms in sample*

Type of company	British companies	West German companies
Private	0	2
Independent public	2	5
Public wholly or partly owned by machine tool group	1	2
Public wholly or partly owned by conglomerate with other interests	4	0
Total	7	9

was considerable evidence of the influence of the founder's family. This appeared to contribute to a greater orientation to engineering than in the British company, where financial considerations predominated (Table 4.1).

Indeed, one of the most pronounced environmental differences evident in the present study was the importance given to engineering skills as a prerequisite for general management. Most of the West German managers interviewed in areas such as marketing, sales and finance, were engineers. By contrast, many of the British managers interviewed in these functional areas did not have the same engineering background.

In the case studies reported in the present study it was clear to the author that the dominance of the engineer in decision-making within the companies had a major influence on the relatively high importance given to product quality or technical performance. It was also clear that the new product development process was more effective because all parties (Sales, Engineering, Finance and Production) were able to talk at the same level about engineering or engineering-related topics.

This finding is consistent with the views of Lawrence (1980). Commenting on several studies of management in West Germany he stated: 'Graduates in business economics do not dominate the commercial functions to anything like the extent that engineers dominate the technical functions. And, not only are the engineers dominant on the technical side, they have also "overspilled" into the commercial function, at all levels, including the highest.'

Such a domination, according to Lawrence, also extends into corporate planning. He commented: 'The contradiction is that German managers extol the capitalist system, certainly with more favour than their British colleagues, but tend to view their companies in technical and quality terms rather than economic performance units.'

In terms of trade associations in the United Kingdom, the machine tool industry is supported by two trade associations, MTIRA (Machine Tool Industry Research Association) and MTTA (Machine Tool Traders Association). MTIRA specialises in technical development work for the industry, whilst MTTA's role is to promote British machine tools generally through the medium of trade shows, overseas exhibitions and commercial information. By contrast, in West Germany the machine tools division of VDMA (Mechanical Engineering Industry Association) helped its members with both technical and commercial problems. Development work within the association seemed to the author to be more easily related to the commercial problems of individual member companies.

As far as universities and other research institutions are concerned, West German technical universities have traditionally played a more active role in product development work than their counterparts in the United Kingdom. Much of the credit for sophisticated equipment developed in West Germany appears to be due to the machine tool faculties of these universities, and their close cooperation with industry. For example, the technical director of one West German company was also a professor of mechanical engineering in the local university.

One of the most active research centres in West Germany is Aachen Technical University. Professor Opitz (Head of the Department of Machine Tools and Production Technology) commented on the relationship between universities and manufacturers in West Germany: 'The only way to success – this is an unavoidable condition for survival in the markets of tomorrow – is an all-embracing coordination of all efforts in the fields of research and development, in which industry, universities and government departments work closely together and join in making available the required financial means.'

In the United Kingdom there have traditionally been far fewer collaborative ventures between the university sector and industry.

Table 4.2. *Export performance of firms in sample*

Country	Percentage of sales in 1978 exported (by value)					
	0–30%	31–40%	41–50%	51–60%	61–70%	70%+
West Germany	—	—	2	6	—	1
United Kingdom	1	3	2	—	—	1

Research within the university sector tends to have concentrated on the areas of invention (fundamental research) rather than application (development research). Only relatively recently have there been initiatives to establish linkages in the areas of research applied to specific company problems (for example, the Science and Engineering Research Council's Teaching Company Scheme). The recent initiative on ASP (Automated Small Batch Production) which became a programme of research and development into flexible manufacturing systems is a similar collaborative venture.

In terms of collaboration between manufacturers, West German respondents in the study spoke of cooperation rather than competition between suppliers. A common view was that it was pointless engaging in direct competition when companies could work together more effectively. Examples of such collaboration included a three year joint venture between two lathe manufacturers to produce a numerically controlled lathe. A second example was an association of four other companies who agreed to design mutually a completely new series of turning machines, to assign a range to each of the member companies to manufacture, and to market the products through a joint sales organisation. There were no examples of such joint ventures in the companies visited in the United Kingdom, even though such cooperation could clearly extend the ability and resources of individual companies.

Export performance and product range

The export trading performance of each company in the sample for 1978 is detailed in Table 4.2. The table clearly indicates that all of the West German companies exported more than 40% of their production, whereas only three out of the British companies did so. The figures are for one year only, but the difference is substantial and

Table 4.3. *Products manufactured by firms in sample*

Product type	Number of West German firms	Number of British firms
Boring	4	4
Broaching	—	1
Drilling	2	4
Gear cutting	2	1
Grinding	4	5
Milling	6	4
Planing	1	1
Sawing	1	—
Screwing/tapping	—	1
Turning	4	5
Machining centre	3	4
Copymilling/engraving	1	—
Spark erosion	1	1
Transfer line	5	5
Honing	—	1

reflects the long term pattern of differences for the industries as a whole.

The one significant British exception to this general pattern, exporting more than 70% of its production, manufactured a specialist product for thread-grinding to high tolerances. This product was manufactured to individual customer requirements, and the company had approximately 25% of the world market for the product in 1978.

Table 4.3 indicates the range of products manufactured by the firms in the sample. The table indicates that the spread of products made by the British and West German companies was similar (and therefore meaningful comparisons could be made).

Detailed comparisons of the level of technology in the individual case studies were not made. Two factors militated against this. Firstly, such comparisons would have been extremely complex and consumed more time than was justified in this research. Secondly, and more importantly, it was felt that the customer's perception of the technical sophistication of the product was more important than the researcher's perception. Detailed comparisons from a customer's point of view are made in the next chapter.

Table 4.4. *Number of years covered by company plan*

	3 years	5 years	No set pattern
West Germany	5	3	1
United Kingdom	—	3	4

General management – planning and control

There were few differences in the approach to general planning and control of the firms in the sample. All had a general management committee which met regularly to control and coordinate the firm's activities, and the same management functions generally speaking were represented on this committee in both samples.

However, there was one major difference. Three of the British companies had a Marketing Director by title on the main board. None of the West German companies had such a director, reflecting the general predominance of engineering management on the controlling boards of the West German companies. This is consistent with the theme which is developed in subsequent sections that the British companies in the sample had many of the trappings of marketing but less of the substance than their German competitors.

Each of the companies in the sample had plans which were rewritten annually. However, there was a marked difference in the time periods covered by the plan, as Table 4.4 shows. Broadly speaking, the West German companies tended to plan to shorter time horizons, exhibiting a more mechanistic or controlled orientation than the British companies in the sample.

The 'no set pattern' category are firms which did not regularly plan the activities of the company formally over a specific future period, or had a flexible company plan with little forward planning.

It is clearly a matter of some concern that at least four of the British companies in the sample appeared to have no set plan. This implies a lack of forward planning about the firm's business and its markets which might in the long term lead to serious problems if new technologies or markets were neglected. Where company plans did exist, there were no obvious differences between West German and British companies. For most companies the major components of the company plan included future sales and profit targets, the introduction of new products and the deletion of old ones, capital investment decisions, expansion into new markets and personnel policies.

One important difference evident in the planning activity was the greater formal involvement of the workforce in the West German companies through the 'Wirtschaftsausschuss' or firm's economic committee. This body, on which management and workforce had equal representation, discussed matters and plans which could have a significant influence on the interests of the employee and had the right to be given information about the firm's economic performance in terms of sales investment and profit. This approach to participation was noticeably different from that in British companies in the sample. For example, a report of the joint shop stewards' committee of T.I. Churchill Ltd (1978) was widely critical of the actions of the management of the group, by inference regarding itself as outside the decision-making process.

This difference may be one of the most important inhibiting factors in the British machine tool industry's attempts at product innovation. From discussion with management in several of the British firms in the sample it was clear that they perceived that the response of the trade unions was one of the major obstacles in introducing new products. They (management) clearly felt that the unions would perceive the introduction of a new product as an opportunity to negotiate new wage rates and working conditions. This appeared to be a major obstacle to overcome in developing and introducing a new product. However, there was no direct evidence to suggest that the trade unions themselves would resist such change, particularly since they perceived a clear link between successful new product development and continued employment.

Indeed, the overall climate appeared to be conducive to the introduction of new products as far as one group of union representatives was concerned. The Coventry Machine Tool Workers' Committee said in 1978 (Joint Shop Stewards' Committee):

> The UK industry's failure to supply British firms with the N.C. machine tools they have been increasingly wanting to install is a direct consequence of the short term strategies the machine tool firms pursued in the sixties. British machine tool firms were most vulnerable to major trade recessions because they were comparatively technically backward – both in terms of the designs of the machine tool

Table 4.5. *Methods of discussing controlling new product development*

Method*	West German companies	British companies
Informally as ideas come forward	9	7
Informally through papers distributed to managers involved	7	5
Product Development Committee	2	4
Management Committee	5	3

*More than one method is possible.

they produced, and of the processes by which they produced them . . . they were, therefore, unable to compete effectively against their chief international rivals.

Organisation for product development

The types of organisation for product development varied considerably in both samples depending on the degree of change. For all companies where product development was of a minor continuous nature (minor changes in machine specification), little or no organisational adjustment was needed. Discontinuous changes on the other hand, such as the introduction of a new type of machine tool, required specific forms of control. For most of the companies in the sample this type of change was managed and coordinated by a product development committee, or general management committee. Table 4.5 indicates the type of organisation adopted by companies in the sample.

There were differences in the form of organisation for new product development. In five of the nine West German companies, new product ideas were discussed formally by the general management committee, and product development committees had some responsibility in only two companies. By contrast product development committees played a more significant role in the British sample.

The pattern of control of the companies in the sample was clearly one of the reasons for this difference. In the majority of the West

German companies control was still exercised by descendants of the founder of the firm. Therefore, product policy was still firmly in the hands of the general management committee. This situation was not typical of the British companies where product policy was more likely to be a devolved responsibility with the general management committee more concerned with financial rather than technical problems.

Project groups, set up to manage the development of a new product idea, were not commonly used by companies in the two samples. Such a group would normally be made up of representatives from different departments within the organisation, and would be formed to manage a development from idea creation to commercialisation (i.e. a form of temporary product development committee). Two British and two West German companies did use this approach, but there were no significant differences in the way in which it was applied.

Approaches to new product development

The following section describes the major differences apparent in the approach to new product development in the companies in the sample. This analysis follows the conventional sequence for new product development activity, namely idea generation, general screening and business analysis, prototype development, test marketing and, finally, commercialisation. This sequence is rarely found in reality in its fully developed form, but it is used here as it provides a useful framework for presentation of the results.

Idea generation – market analysis

Six of the seven British companies in the sample had formal Marketing Departments, whereas only four of the nine West German companies had such departments. Where companies had such departments, the responsibilities of the department were broadly similar, covering publicity, market research, pricing and product development. As an earlier section has commented, marketing was only formally represented at board level in three companies, all of which were British.

In the one British company without a Marketing Department, the activities of publicity, market research, pricing and product development were controlled by the general management committee. By contrast, in the West German companies without Marketing Departments, these activities were shared between other departments. For example, in company C (German), the company's Technical Director described the approach thus:

> Marketing of the company's products is a combination between the activities of the Sales Department and Technical Department who continue to decide on new products, analyse the competitive situation, fix market prices and fulfil all other marketing functions. There is no fixed organisation responsible for the marketing programme, it is part of the responsibilities of the Design Department and the Sales Department.

On closer examination, whilst the British companies were more likely to have a formal specialised activity (termed marketing) responsible for the analysis of customer needs and new product development, the West German companies were more likely to have direct contact between the actual design and development activity and potential customers. This tendency to create a specialist function appears to be a general characteristic of British companies. Lawrence (1980) finds a similar difference in his studies of West German production management. He commented:

> The British companies tend to 'externalise' specialised knowledge at all levels from the shop floor upwards. They tend to create ancillary technical and other posts external to the main body of production workers and to line management hierarchy. The German tendency, in contrast, is to incorporate such specialist and technical knowledge in the line itself and to expect production workers and their immediate supervisors to do as much as possible for themselves rather than depend on staff specialists and others outside the line.

The major source of ideas for new product development work was the trip report of the sales engineer. Analysis of reasons for lost orders or salesmen's comments on the technical suitability of current products compared to competitive ones and customer needs often gave rise to ideas for development.

However, as Table 4.6 indicates, the analysis of sales trip reports

Table 4.6. *Managers responsible for analysis of sales trip reports*

	Marketing	Sales	Design and Development
West Germany	3	5	4
United Kingdom	6	3	—

was more likely to be done by the design and development engineers in West Germany. Such engineers made a direct evaluation and review of these reports, whereas in the British companies the Marketing Department or Sales Office frequently made this evaluation which they then passed on to the Design Department. Marketing or Sales were therefore an intermediary between the customers' needs and the design function, interpreting the information for the Design Engineer.

The evidence of the case studies suggests that this may well reduce the efficiency with which the design engineers in the British companies could interpret customer requirements and respond to the needs of the market.

The design engineers in the West German companies were also more likely to visit customers than their British counterparts. For example, in company C (German), the Design Department was organised by product type, and the head of each Design Group was normally the top specialist in the company for the machine type. Each of these group 'heads' normally had close and continuous contact with the company's major customers. (One estimated that he spent between 20% and 30% of his time visiting customers.)

By comparison, in company J (British), contact between the Design and Development Department and the market place was indirect, in the sense that other departments in the firm provided market information and fed it to the engineers. This company's Design and Development Manager commented:

> New developments tend to come about because of observed developments of the competition or the Board's prompting. Our own scanning takes two basic forms, reaction to competitive sales literature, and watching what's happening in a general sense – for example, developments in microprocessor technology. We tend to

> make a general use of information, there is no system as such – managers spot things and feed them through to us.

Ad hoc market research, in the form of a special one-off study of the market, was rarely encountered in either country. (There were only a few examples of specific market research exercises to identify 'gaps' in the market, i.e. needs that were not currently being satisfied.) In company C (German) this was a relatively limited exercise aimed at identifying reasons for not achieving sales targets. The company's Design Director commented:

> We try to find out why we didn't meet our target market share – wherher or not this was due to technical reasons or price reasons – we go back to customers and ask them – usually it's an arrangement between the sales manager and myself, we would visit three or four of our main customers and discuss the current machine and its problems – we also have a good feedback from our fitters who ask people informally on the shop floor if they have trouble with competitors' machines, etc.

In company J (British) market research to generate new product ideas played a minimal role. The company's Marketing Manager commented:

> There is no need to review sales reports, we use lost order analysis, sending out questionnaires to our salesmen – that's about all we do . . . the marketing input into new products has been minimal until recently. In this company there has traditionally been a resistance to using market research for new product development.

The majority of respondents in both countries appeared to believe that market research had only a limited usefulness in generating new ideas. The general feeling was that customers were more likely to respond to specific proposals, rather than initiate such proposals themselves. One manager's comment was typical of a general view. He said: 'I don't think customers have enough far-sighted vision to specify new machines in isolation – to be in constant communication irrespective of specific demand. We only hear from them when they have specific needs.'

Only one company (company L, British) produced general market studies on an *ad hoc* basis for individual operating companies within the group. The group market research function carried out such studies. In the words of the group Marketing Manager: 'This type of review would indicate changes in the specification and technology required to stay in the market place – to act as a stimulus to the manufacturing divisions, we try to sound warnings of the future, whether or not a product should be changed, etc.'

These reports were then fed to operating management at divisional level, and the implications for product development discussed. However, the major problem identified was 'getting operating and design managers to take any notice'.

Idea generation – research and development

From discussions with the firms in the sample, it was clear that the majority of development work in the machine tool manufacturing companies was continuous improvement in the specification of existing tools involving only minor changes in the tools themselves. Many of the ideas for discontinuous change tended to come from outside the industry. For example, developments in cutting tools which enabled higher metal-removal rates, control systems with new capabilities or development in work-handling systems, automated assembly or manufacture.

It was clear that there were several differences between British and West German companies in the quality and extent of their own research and development activity. The point has already been made that West German companies appeared to be involved to a greater extent with universities and other research institutions. It was also clear that the technical function had greater involvement with customers on a continuing basis.

A third major difference between the nature of design and development activities was that the West German companies in the sample tended to coordinate these activities in one central research and development function. By comparison, in several of the British companies in the sample, the research and development activity was conducted at divisional level, rather than on a group basis. In effect, therefore, British companies lost any advantage which might have accrued from centralisation (for example, the resources to employ specialists, and the extent to which development work in different areas could be complementary).

A further result of the greater centralisation in West German companies was that the technical group involved in research design and development tended to be larger than in the British counterpart. For example, company F (German) employed approximately 200 staff in research, design and development, and company G (German) employed approximately 160 staff in the same function – the corresponding figure for company O (British) making products directly competitive with companies F and G was 20. Company O was part of a group of machine tool manufacturing companies, each of which had its own separate research design and development group, rather than one integrated function for the group as a whole.

Comparisons between new product development techniques

Table 4.7 details the approaches to the initial screening and business analysis of a product idea adopted by the companies in the sample. The descriptions appearing in this table are based on discussions with managers in each company, and are related to specific recent developments within those companies.

Each company appeared to go through several stages in the initial screening and business evaluation of new product ideas. The first of these stages can be labelled 'establishing initial feasibility'. A feature of the West German companies in the sample was the extent to which they tended to involve the customer at this stage much more than their British counterparts. The German company was more likely to take the original design concept to two or three chosen 'friends' or customer companies to establish their initial reaction.

On the other hand, the attitude of British supply companies appeared to be likely to preclude an early reference to customers on an informal basis. In some of the British companies in the sample, it was felt to be of little use to refer directly to customers at the early stages in development because the customers themselves were incapable of giving a meaningful response to a general product concept, and were unwilling to try anything new. Indeed, in one company (company J) it was felt that if a customer became aware of a new development then this might put him off buying from the existing range until that development was available on the market.

Significantly, perhaps, the major contact which occurred between British firms and their customers at the stage of establish-

Table 4.7. *Initial screening and business analysis*

West German companies
Company A Engineers in Development Department produce first drawings from original idea and present to Product Development Committee. If suitable in general terms to company's business (unstructured evaluation), Project Manager appointed and development proceeds. Lay-out shown to 'a few customers – not big shops, but people who are interested in improving ideas – in new technological solutions, we rely on this close acquaintance'. At general assessment stage orientation is to world market. 'We can't look at individual markets because West Germany is too small to plan new products for it exclusively.' Production Engineering estimate manufacturing costs, market price and sales volume are estimated as is production loading. Results of this study presented to company's Board for go–no-go decision.
Company B Product Manager in charge of product produces proposal to product planning group, indicating the general problem and customer needs. If proposal is accepted, passed to technical group who produce first design in outline form only. Product Manager then presents design to selected customers, and produces a forecast of the likely demand and market price. If product planning group accept figures further design work is done on machine, using a team of value engineers within the company to get manufacturing costs as low as possible, until a final design is 'frozen'. Go-ahead to build then given if machine has an acceptable profitability at the forecast market price and there are no better alternatives.
Company C Original idea from Design or Sales taken informally to two or three customers and reviewed with them. Following this informal discussion, idea discussed in planning meetings, to provide rough estimates of cost, time and market potential. Key problem is to establish 'going market price'. Design group then evaluate and modify design to meet this price at a profit. Separate value engineering group not involved in original design, examines alternative machining techniques and assembly methods. Extent of customer contact varies considerably, depending on degree of innovation in product, and company's own experience with similar technologies. 'We know perhaps 80% of these conditions very accurately, but to get the remaining 20% we have to talk carefully with our customers about their needs. These 20% may form an option and not be part of the

Table 4.7 (*cont.*)

basic machine, but it is necessary to develop all the options which will be needed at the beginning.'

Once design 'frozen', go–no-go decision taken – based on a comparison of return on cost of development incurred and forecast over anticipated life cycle of product, with return from any other development project currently under way. In periods of scarcity project with highest expected return preferred.

Company D
Original idea tried out on a few chosen customers for opinions on price and product attributes. Product attributes determined in discussion then costed and total cost compared with machines already on market. Where change necessary entire design vetted by independent design analysis team who examine materials and manufacturing costs. Go–no-go decision depends ultimately on profitability.

Company E
If an idea appears promising a desk-based market study is done, detailing competition and their technology and assessing company's own prospects. Technical aspects of product discussed between Sales and Design and product developed. Product idea discussed with customers and modified where necessary.

Company F
Machines are purpose built to reflect individual customer needs. Process is a series of *ad hoc* adjustments of the basic product which is not systematic. Success of adjustment depends on ability of company's sales force and design team to interpret customer requirements at a profit.

Company G
Ideas from sales force or Design are discussed in a general manner at the product planning meeting. Given general agreement Sales Department will approach customers to ask them about general concept. Key stage here is identifying 'friends' or 'promoters' who are receptive to innovation in the company.

Fixed design is evolved from customer feedback, and costed. Independent firm of value engineering consultants used to reduce manufacturing costs as much as possible.

Final costs are compared with going market price, and anticipated life cycle of product. Go–no-go decision depends on anticipated return compared with return from other projects.

Table 4.7 (*cont.*)

Company H

Original idea generated in long term planning session, and given to an engineer as a design project. Engineer collects more data and reports back to a future planning session.

If project seems initially promising, engineer asked to produce a general design outline, and sales engineers asked. They form opinions by asking certain key customers.

If sales feedback is favourable, then detailed budgets are evolved by asking engineering, workshops, and sales to provide costings and five year sales forecast at a price based on preliminary costings.

Development budget and sales forecast presented to management committee who then decide on development – based on return-on-investment criteria – developed costs must be paid back within two years.

Company I

Individual product built to individual customer requirements – each project judged on its own merits.

British companies

Company J

Board initiates a request for new machine, indicating in general terms product type and selling price. Design and Development produce first design, and Sales Department asked to produce estimates of market potential. Customer not directly consulted at this point because 'if they get wind of a new product in the offing then they will wait for the new product, and you lose orders. We don't do a direct questionnaire survey of customers' needs because they will ask for all sorts of different things – usually with confusing results.'

Detailed designs and estimates of sales potential are then returned to the Board for judgement and approval.

Company K

Ideas for new product development can originate from Sales (most frequently) or Design (less frequently). Where substantial change is involved Engineering and/or Research and Development will produce a draft design and submit to the Product Development Committee.

A formal check list of items was once used to evaluate each project (five years previously). Now system is less rigid focussing on compatibility of project with company's activities and capabilities, production resources needed, degree to which it meets the market's needs, and development costs. Go–no-go decision depends on how Product Development Committee assess project on these criteria. If given, development proceeds without direct reference to customers.

Table 4.7 (*cont.*)

Company L
Idea for product development originates from individual customer contacts (or less frequently market research reports from group). If idea substantial project group may be formed. First outline design is prepared by design team in manufacturing company. Customers are rarely directly involved in development at this stage.

Specifications are discussed by production and design management, the key criteria being whether the product can meet a target market price at a high enough level of demand, and Group Marketing Services assist in market evaluation.

Go–no-go decision is taken at divisional level based on this overall assessment.

Company M
No systematic procedure to evaluate new product development because most development work is based around needs of particular customers. 'If the sales engineer on a visit to a customer saw that something was needed, and we can see it would enhance our product range, then we would try to develop around the customer . . . this tends to be the development of our existing machines – very rarely is that something new.'

With discontinuous innovation there is no formal assessment, simply the opinions of the Managing Director and the Sales Manager. 'If these two are in favour then it usually means that the idea has a fair amount of validity . . . and it is really taken as that . . . we don't really feel it is justifiable in doing any further analysis.' The company tends to proceed on the basis of this 'feel' for the market.

A draft proposal is usually prepared for major developments, with a brief analysis of the costs in developing an idea and manufacturing the finished item, but this does not influence the go–no-go decision as much as the personality of the idea sponsor. 'Really the next step is so much dependent on the individual – if he is fairly strong-minded – if the idea is good, then it will move at a much faster pace – but it is up to the individual – if he feels strongly about it then he will be given the opportunity.'

Company N
Much of the discretion for the original proposal is left with individual plants and the Plant Managing Director has a degree of autonomy in the products he supports.

If the project is agreed by the Product Review Committee then detailed design work will be done, and drawings will be produced.

Proposals for new products in company N have to be submitted formally to the Product Review Committee and have to provide detailed infor-

Table 4.7 (*cont.*)

mation on the product features needed by market, the proposed market size, cost to produce the machine, and anticipated market price.

Customer involvement in this process tends to be minimal 'because it's very rare that the customer is going to originate the new concept – for example, if it were left to the customers we still would not have certain technologies. However, there are some customers who are worth listening to – they tend to be the companies with specialists whose job it is to think ahead and decide what the production requirements are going to be.'

Formal go–no-go decisions on major developments are given by the company's main Board, where 'if they are in any doubt they will ask for further information – with a Managing Director who is an accountant this is frequently financial.'

Company O

There is no clear pattern for the majority of product development in company O because a great deal of the company's output is special purpose machines built to individual customer specification

With major discontinuous innovation (rare) initial product designs are submitted to the product review committee for evaluation and comment. Marketing may evaluate the proposal, doing a specific market research study, but 'in many cases we wouldn't do market research because it's a small and well-known market'.

Value engineering is not usually carried out on the design because the small potential savings it is felt could be achieved would not be justified by the extra costs of the activity, especially in a market which was price-inelastic.

Company P

Proposals on new product development from Sales/Marketing and Design and Development are submitted to the Product Development Committee. Continuous modifications of the minor variety require little formal justification, but the design and construction of special purpose machine tools to an individual customer's specification is subject to detailed examination on a case-by-case basis.

Decisions on the development of a completely new type of machine tool are evaluated using a formal check list, where the most important criteria are the support which the new product idea gives to existing products, the pay-back period, the development costs, manufacturing requirements, and the launch cost.

The go–no-go decision is given by the Product Development Committee, based on its appraisal of the proposal.

ing the initial feasibility for the product idea, if contact took place at all, was likely to be formal market research of customers and their needs. This approach, of which several respondents were cautious, was singularly absent from the approach adopted by the German companies in the sample.

The West German companies appeared to have a higher degree of contact with a few carefully chosen customers at the initial stages of establishing the development's feasibility, and relied to a limited extent, if at all, on formal market research. The greater involvement of the design engineer in the West German companies in the sample, already mentioned in several places above, was a major differentiating feature of this early stage of establishing the development's initial feasibility.

The second stage in the initial screening and business evaluation process can be labelled 'preparation of detailed case'. Reviewing Table 4.7 there were no significant differences between British and West German companies in the sophistication with which the detailed case for each development proposal was prepared. In both samples estimates of sales potential, likely market price, development costs, and production and launch costs were generally required before a decision could be taken on whether or not to proceed.

A significant difference was evident in the way that the original design was modified and refined. In the West German companies in the sample the cost-effectiveness of the design was more likely to be checked by an independent value engineering team than it was in the British companies, where the efficiency of the design was more likely to be left to the original design engineer. Independent scrutiny of the design by a separate value engineering function, either within the firm operating as an independent unit to vet individual project proposals, or in the form of outside consultants called in to assess individual design alternatives, would be likely to give more attention to the manufacturing costs and design efficiency of alternative development proposals.

Where a systematic value engineering procedure was used this would seem to be more likely to result in products representing 'value for money' than where such a procedure was not adopted. The difference goes part way therefore to suggesting why West German products from such companies may represent better 'value for money' than those from their British competitors.

Table 4.8. *Approach to prototype development*

	West German companies		British companies	
Develop prototype machines	Yes	6	Yes	5
	No	1	No	—
	Rarely	2	Rarely	2
Location of prototype testing	In house	—	In house	4
	In customer's factory	6	In customer's factory	1
Duration of prototype testing	0–3 months	—	0–3 months	1
	3–6 months	—	3–6 months	1
	6 months–1 year	6	6 months–1 year	3

Prototype development

Once the original development idea had been sanctioned, the next stage in the development process was to produce detailed drawings and ultimately to manufacture a prototype of the product development in question. Clearly, for many minor continuous changes in product specification the construction of a prototype machine tool was not required. However, where new types of machine tool were concerned, prototype machines were often developed. Six of the West German companies in the sample developed prototype machines regularly for new types of standard machine tool, and five of the British companies in the sample also developed prototype machine tools for the same type of development. (Prototypes were obviously not built by those companies which specialised in special-purpose machine tools to individual customer specifications, since the machine itself was the prototype.)

But there was a major difference in the way in which the prototypes were tested. All of the West German companies tested prototypes in the works of one of their customers. On the other hand, the British companies in the sample were significantly more likely to test the prototype 'in house'. Only one of the British companies in the sample which regularly developed prototypes for new types of machine tools said that they tested these prototypes in the customer's factory. Moreover, as Table 4.8 shows, the West German companies tended on average to test prototypes for a longer period than British companies in the sample.

There appeared to be a fundamental difference in attitude between the two samples of companies as to the role that the user organisation can play. Managers in the West German companies spoke generally about the benefits of having a user who could be relied upon to give any new development a thorough test, and who would be receptive to trying out a new machine despite any obvious problems which such a prototype might have.

For example, a comment was made in company D (German) that: 'We try to get a customer who is a good friend and close to us, where we can do a quick service and check the machine daily where necessary. We should have a good relationship with them.'

Asked to define what sort of customer this would be, the same respondent indicated that this customer would be probably close at hand, not necessarily highly technically qualified, and would be capable of not saying anything about the development until it was proven.

In company E customer involvement was viewed as essential 'because the machine is never completely right from the word go'. Speaking of customer involvement in the prototype development stage, one respondent from this company stated: 'We always change who we go to with new machines – for some machines one manufacturer is the right one, and for others another is right. However, in some instances it is difficult to get good relationships with particular companies.'

Management in company G (German) spoke in terms of having certain 'friends' or 'promoters' in user companies who were receptive to innovation. He commented: 'They want innovation because it is in their interests to be seen to be using leading techniques of production – they want the feeling that they have the best production equipment. This is particularly the case in the motor industry.'

By contrast, one respondent from company J (British) spoke of the problems of using customers to evaluate prototypes. He commented:

> Customers don't want to have to solve all the problems in a machine, and to do this [extended trial of the machine in the works] you need a customer who is sympathetic with your company and understands your problems . . . You need someone who is ready to stick his neck out, and there are not many of these in British manufacturing industry . . . Certainly, not amongst our customers.

A respondent from company L (British) commented in a similar way about the problems of involving a domestic customer:

> In the United Kingdom one of the biggest problems is the reluctance of the customers to get involved in development of new machines – there is an inherent conservatism of the customer and a reluctance to accept the responsibility of being the first – the customer is rarely involved in the design of 'standards' – he wouldn't usually be involved until the machine was shown with dramatic effect at its first exhibition.

In company M (British) the attitude was expressed that the customers' opinions were relevant, but that the customer was not likely to be involved directly in the development of new machines. One respondent commented:

> We certainly have major customer types who buy a lot of our machines but they will not necessarily be involved in anything new we are doing – we would tell them about it and look on their response with interest . . . They may have some influence if they come along and say why are you doing that? . . . We may rethink if we value the company's judgement and the person's judgement, again so much rests on the individuals involved rather than the system.

The cost of building and testing of prototypes was also used by one British company to justify the relative infrequency with which prototypes were developed. In company O (British) the marketing manager commented: 'If you have a machine that is proven and run, then you would get more customers, but to do this you would have to run a considerable number of components through the machine and this costs time and money.'

Differences in the suppliers' attitudes to customer involvement in prototype development and testing are clearly one of the significant variables in this study. The quality and extent of interaction between user and supplier, certainly from the supplier's perspective, appears to be greater in the West German companies in the sample, and perceived as a matter of course rather than the exception. Chapter 5 takes up the issue of the involvement of customers in the development process in greater detail, and therefore more extended discussion of the nature of customer involvement in the development process is reserved until then.

Criteria important in product design

The starting point for this study was the observation that differences in technical quality of the products offered by British and West German companies may be one of the main reasons for differences in their comparative international trading performances. Broadly speaking it was hypothesised that West German companies had given greater attention to the technical quality of their products rather than to price as the major dimension of competition. The following section of this chapter discusses the approach adopted by companies in the sample to determining product features. The discussion also considers the relative importance given to different product attributes by companies in each country.

Management in companies in both countries emphasised the trade-off which was necessary between achieving economies using a high degree of modular design where possible, and increasing the attractiveness of the product by 'tailor-making' the machine to individual customer requirements. In this industry the main elements of modular design were items such as the tables, knees, drive systems and motors used in different machine tools.

The process was summarised for one company (company H, German) as follows:

> The problem is to find the right combination of product features to meet perhaps 70% of the customer's needs. A 100% machine would be too costly to build, and so we have to compromise at say 70% of the needed features, and then add on special features to meet individual customer requirements. In design terms we have to think about three distinct sets of criteria – those which we must have, those which we would have, if possible, and those which it would be nice to have.

The 'matrix design' approach involved a formal assessment of all the potential attributes of a new machine tool in order to produce a specification for the '100% machine'. These attributes were listed by comparing customer requirements with technical capability. Individual attributes were then successively eliminated by a process of value engineering, and reference to key users, until the machine's performance and cost to build bore a satisfactory relationship to a target market price. If the market price could not be reached satisfactorily then the project was abandoned.

Despite the emphasis on market price which was clear in most of the comments of West German managers in the sample, it was also clear that the firms involved had reconciled themselves to not being competitive in the production of standard types of machine tools. Several comments indicated that the continuing high value of the Deutschmark in world currency markets meant that the manufacturers in West Germany had to produce machine tools which were special purpose or designed for individual requirements. High technology products, rather than standard products, had become the norm in remaining competitive. Other essentially 'non-price' factors were also important. In particular, delivery and back-up service were also important. One manager in company F (German) commented on the relative importance of price and product quality in competitiveness.

> Today the main problem in competition is price, but in addition to what we can show and demonstrate, there are other things which the customer would want to see perhaps unconsciously – we feel that the so-called unquantifiable aspects are important, one of them is the buyer's view of the potential life of the machine – but how can you quantify this? . . . It is 99% the reliability of the product which creates good relationships – quality in terms of how efficiently the machine works, and then reliability.

Product quality had also become the main basis of competition in company G, where one manager commented: 'In the UK, for example, we can only sell by solving bigger problems than the usual ones – it's not usual to sell standard machines in Great Britain.'

A different philosophy, however, was evident in the United Kingdom. For example, in company K price was perceived as the most important factor. The company's Chairman commented: 'We are in the marketplace where our products may be measured against others similar to our own, and we must continue to provide quality and value for money in an endeavour to obtain a maximum share of available orders'.

The company's Design Director also emphasised this point. He commented: 'The selling price is crucial – as soon as the specification is written for the machine that is where the selling price begins to accumulate. Marketing keeps us on the straight and narrow, the selling price is all we specify – all they want is the maximum specification to the lowest possible price.'

Of the British companies, only company K appears to have been successful in competing in volume markets with a standard price competitive product.

Company N, the largest British manufacturer of machine tools in the sample, had continued to produce relatively standard products, and had attempted to sell them on a price-competitive basis, predominantly to buyers in the developing economies, particularly those of the old Commonwealth. As these countries were now able to produce their own machine tools, company N was having to readjust to producing sophisticated products more in line with specialist user needs. Company L had similar problems with its own products but the reasons for its situation were different.

Company L (British) described its philosophy as attempting to 'round-out' its range of standard machines with imported machine tools from overseas manufacturers (some of which were German). As a result of this policy the sophistication of its own machine tools seems to have developed only marginally, since it was able to satisfy its customer needs by acting as an agent for the overseas manufacturer. In the long term such an approach could only be damaging to the domestic production of more sophisticated products.

The evidence of the case studies supports the overall hypothesis that product quality or technical sophistication was more likely to be stressed in the West German companies than the British companies in the sample. Macro-statistics on the value per ton of British and West German machine tool manufacturers and the comments from the case studies support this particular viewpoint.

Concluding remarks

Chapter 4 has examined product development activities in a sample of British and West German machine tool manufacturing companies. From the evidence presented in this chapter it is clear that there are significant differences in philosophy and approach. Whilst such a broad cross-national comparison may hide many similarities in product development the author believes that certain general patterns have emerged from this review which have an important role in explaining differences in success in product development.

However, this picture is not complete since it ignores the role of the customer. From this chapter it should be evident to the reader that the customer has an important role to play in the product development process, and the next chapter turns to an examination of this role in the markets for British and West German machine tools.

5

The role of the customer in new product development

Introduction

Studies of the process of technological innovation have traditionally been supplier-orientated, in that they have concentrated on characteristics of the supply company and its approach to new product development. The study of suppliers reported in the previous chapter is typical of such an approach. This type of study can frequently yield useful generalisations about ways of improving an organisation's approach to new product development. However, a study of the approach of the supplier to technological innovation is deficient to the extent that it ignores the potential contribution of the customer.

There is already some evidence in the literature, supported by the results of the last chapter, that the customer may frequently play an active role in initiating change in a supplier's products, and in influencing the development of product changes made by the supplier. The last chapter has shown how West German suppliers in the machine tool industry tended to involve their domestic customers more directly with the machine tool development process than their British competitors. This reflected a more positive attitude on the part of West German manufacturers to customer involvement. It may also have reflected a more positive attitude towards technical change amongst West German customers. They may have been more active in demanding change from their suppliers and receptive to innovations which were introduced. In such a climate it might be expected that the West German supplier, with a more demanding and receptive domestic customer, would be under greater pressure to continue to develop improvements in the technical quality of its products. This possibility is examined in this chapter.

Before considering the results of the study in detail the focus on the domestic rather than overseas customer as the reference point for suppliers in the process of technological innovation should be justified. Clearly it could be argued that suppliers in West Germany and the United Kingdom have the same access to all potential customers whether domestic or overseas. However, in discussion with the West German suppliers, it became clear that they tended to use the domestic customer much more extensively than the overseas one in product development work. Managers spoke of having a customer who was close at hand, providing quick feedback on new ideas. Problems of language and geographical distance tended to preclude reference to overseas customers. On the few occasions when British suppliers referred to customer contact, they too tended to discuss the inputs of domestic rather than overseas customers, again for reasons of ease of contact and communication, particularly when actually developing the product ideas to the prototype stage.

Differences in the needs of British and West German users

The relative importance of technical and economic factors in the machine tool purchasing decision is at the centre of this enquiry. If the customer plays an active role in the development decision then the importance given to different product features such as reliability, cost and delivery is clearly crucial information for a supplier. These features can be distinguished from other essentially 'supplier-related' factors such as aftersales service which are not specific to an individual product.

Table 5.1 illustrates the overall rating of importance given to a series of machine tool attributes by respondents from both countries. Clearly the rating of an individual machine attribute could vary depending on the use to which the machine tool was put, and the individual circumstances of the purchasing organisation. For a specific purchase by a single firm the combination of factors would be likely to be unique, reflecting the requirements of the individual organisation in a particular situation.

For example, if the respondent was evaluating a production machine to be used in continuous large-scale manufacturing, then

Table 5.1. *Overall rating of product attributes*

Question: How important do you believe the following factors are when you are considering the purchase of a new machine tool of relatively standard design?

	Rating						
	Very important				Not important		
Product attribute	1	3	5	7	9	$\bar{x}$	n
Accuracy	76*	23	0.8	0	0	1.49	121
Reliability	91	9	0	0	0	1.18	125
Flexibility	33	39	23	4	0	2.97	120
Price	33	37	25	5	0	3.03	125
Delivery	33	40	24	3	0	2.94	118
Power	36	28	25	9	1	3.21	116
Maximum size of workpiece	23	40	24	7	5	3.6	116
Rigidity	48	36	14	1	1	2.41	115
Standardisation of parts	30	36	23	10	2	3.37	118
Technical sophistication	36	26	26	10	3	3.35	117
Speed of operation	53	29	16	2	1	2.37	115

$\bar{x}$ is the mean value given to each attribute on the assumption that the scale has interval properties.

n is the number of respondents who provided a rating for the attribute.

* To be read: 76% of the respondents to this question believed that accuracy was an important factor rating 1 on a scale from '1: very important' to '9: not important', when they were considering the purchase of a new machine tool.

the attributes reliability, rigidity and speed of operation would be likely to be most important. On the other hand, if the respondent was evaluating a workshop machine which would be used in a toolroom on a variety of applications, then the attributes flexibility, accuracy, and maximum size of workpiece could have been expected to be of greater importance. Acknowledging these problems of interpretation the question still allows a general comparison to be made between the customers' perception of the relative importance of technical and economic factors. In this respect the study is significantly different from earlier ones which have tended to take the viewpoint of the researcher and define important innovations in terms of the researcher's own perception of importance rather than the users'.

From Table 5.1 it is clear that some technical attributes of the machine including accuracy, reliability, rigidity and speed of operation were seen as important by the majority of respondents. Other technical attributes including flexibility, technical sophistication, power, maximum size of workpiece and standardisation of parts were viewed with varying degrees of importance by different respondents, some rating these attributes highly, others less so. Price and delivery, the two economic product attributes included in this listing, were also viewed with varying degrees of importance.

Another potential economic attribute, operating costs, was deliberately excluded from the list of product attributes which were specified explicitly because of the interpretation difficulties apparent when the questionnaire was piloted. However, an 'others' category in the listing in the questionnaire was included to catch any respondent who wanted to cite this or any other factor as important. There were no significant additions to the original list in Table 5.1.

The results of this analysis are consistent with those reported in the broadly based study of investment decisions in British manufacturing industry cited in Chapter 2. It is apparent that technical product attributes are rated more importantly than economic ones by the majority of respondents. In particular, reliability, accuracy and flexibility are rated more importantly than price.

From discussions with individual machine tool users it was clear that this outcome was consistent with a model of the machine tool purchasing decision which stressed the importance of the ability of a machine tool to do the work required, rather than the importance of price. Price becomes a determinant factor in choice only when the suitability of the machine to do the work has been proven.

This data suggests that a decision to concentrate on price alone rather than on considering improvements in product quality too as the basis for competition will lead to problems in this market. Respondents from both countries gave a greater importance to technical product attributes when they were considering the purchase of a new machine tool of relatively standard design.

The traditional emphasis given to price competition by the British companies in the sample is therefore questionable. From this table greater emphasis on technical product attributes is called for. This data in itself supports the view that British suppliers in this

Table 5.2. *Importance rating given to product attributes 'power', 'rigidity' and 'technical sophistication'*

Product attribute	Country	Rating: Very important 1	3	5	7	Not important 9	Totals
Power†	West Germany	60	26	14	—	—	100 (50)
	United Kingdom	18	30	33	18	—	100 (66)
Rigidity*	West Germany	60	33	8	—	—	100 (49)
	United Kingdom	40	40	21	—	—	100 (66)
Technical† sophistication	West Germany	69	22	10	—	—	100 (51)
	United Kingdom	11	29	38	23	—	100 (66)

*Differences significant at 95% level of confidence.
†Differences significant at 99% level of confidence.
Percentages may not total 100 because of rounding.

industry are not as competent as their overseas competitors in meeting the needs of the market since they have traditionally concentrated on the price variable.

Cross-classifying the rating of these product attributes by country, only three were rated significantly differently by respondents in West Germany and the United Kingdom. These were the power of the machine, its rigidity and its technical sophistication, all of which are technical product attributes. Table 5.2 illustrates these differences.

This comparison shows that the West German customers in this sample viewed these technical product attributes as significantly more important than their British counterparts. The difference in the importance rating given to the attribute 'technical sophistication' is particularly significant. West German customers were more likely to view this factor as important in their evaluation of machine tools, supporting the hypothesis that they would be more receptive to technological innovation in the machine tools which they bought (and more demanding of it from their suppliers).

This result is only one broad indicator of differences between the British and West German customers in the sample, and interpretation of the factors is clearly subject to some debate. However, it is the first indicator of a major difference in the orientation of the

British companies in the sample to new technology, and consistent with findings in later sections of this chapter.

Organisational differences

The decision to invest in a new machine tool can be seen as an organisational problem controlled and constrained by operating procedures, control systems and patterns of allocated responsibilities. In Chapter 3 it was suggested that the most receptive organisations would be those which exhibited the characteristics of the 'organic' organisation, which were relatively flexible and capable of reacting quickly to change.

The West German customers in the sample were significantly more likely to be independently owned companies than those in the United Kingdom, mirroring the distribution of ownership and control in the population of British and West German companies as a whole. From discussion with these companies it was clear that this factor in itself had a major impact on investment decision taking.

In the British companies in the sample which were part of a larger group of associated companies, the individual company's performance had to be justified regularly to the group usually in financial terms. Accountancy data was used to provide the basis of control. By contrast, in the West German companies in the sample, the company was more likely to be an independent company able to set its own goals, and free to use technical as well as financial criteria in setting targets and measuring performance. Accountants were less likely to be in final control, particularly of new investment decisions.

Where financial control is more stringent it can inhibit the rapid acceptance of a new technology where the ultimate benefits, initially at least, are difficult to determine. More conventional technologies with established costs and pay-back periods may be easier to justify to a group financial controller. This appears to be one of the major contributory factors explaining why price rather than technical sophistication was rated more significantly by the British user companies in the sample. A detailed comparison of the machine tool purchasing process in subsequent sections of this chapter supports this view.

There was some evidence of a greater degree of formality in the planning and control of West German companies in the sample.

They were significantly more likely to plan over longer time horizons than the British companies in the sample, again perhaps reflecting patterns of ownership. Items in the company plan were similar in both countries and the plans were updated with the same annual frequency in most companies. The West German companies also held management committee meetings at board level more frequently than their British counterparts, consistent with the concept of collective responsibility for corporate decisions which appears to characterise decision taking in the Boards of West German companies (Lawrence, 1980).

However, at an individual manager level there were no significant differences in the frequency of review of performance of the individual managers replying to the questionnaire. For most managers in the sample their performance and that of their departments were reviewed on a monthly basis. Therefore there was some evidence that the West German company tended to be more mechanistic than its British counterpart, although the differences identified point less perhaps to a resistance to change, than to a better planned and more carefully controlled organisation with greater shared responsibilities.

Organisation of machine tool purchasing

A relationship between the pattern of ownership and control in the companies in the sample, and the organisation of machine tool purchasing, is apparent in Table 5.3.

The table shows greater reference to the parent company in the British companies in the sample (although the difference between British and West German companies is not significantly great). More interestingly it shows significantly greater use of committees or groups of managers to control the investment decision in the British companies in the sample. More managers appear to be involved in the purchasing process, corresponding to an organic rather than mechanistic approach. An analysis of the numbers involved in each country supports this view (Table 5.4).

The numbers involved in decisions on new investment are clearly significant. In 21% of the British companies more than five managers were involved in the decision. From the suppliers' point of view each of these would have to be persuaded that the potential innovation had merit over conventional alternatives. More import-

Table 5.3. *Patterns of control of machine tool purchasing decisions – United Kingdom and West Germany compared*

Question: How is the purchase of machine tools controlled in your company?

Type of control	West Germany Yes	United Kingdom Yes
Decisions are referred to a parent company committee set up specifically to control capital investment decisions for the group as a whole.	19%*	23%
Decisions are referred to a general management committee of the parent company.	25%	28%
Decisions are taken by a committee within your own company set up to make capital investment decisions for the company.	17%	35%
Decisions are taken on a machine by machine basis by a group of managers who meet informally.	23%	36%
Decisions are taken by one manager in your own company.	53%	24%
Base	56	73

*To be read: In 19% of the West German companies decisions were referred to a parent company committee set up specifically to control capital investment decisions for the group as a whole.

antly, with more people involved, the decision process would take longer. For the supplier developing a new product it would be harder to identify the needs of the organisation precisely and meet the requirements of all those involved. Where there are fewer decision takers the decision or feedback would be likely to be quicker. An organic organisation with shared responsibilities may be more difficult to work with and take longer to reach a decision.

During the preparation of the buying company case histories several key stages were identified in the machine tool purchasing

Table 5.4. *Number of managers involved in purchasing process*

Number involved	West Germany	United Kingdom
One	11% (6)	1% (1)
Two	24% (13)	18% (12)
Three	39% (21)	49% (33)
Four	15% (8)	10% (7)
Five or more	11% (6)	21% (14)
Totals	100% (54)	100% (67)

Chi-square = 9.47. Significant at 90% level of confidence.
Percentages may not total 100 because of rounding.

process. These were consistent with those identified in other studies of this general type. Table 5.5 compares the extent of involvement of managers from different functions at different stages in the process.

In Table 5.5 the individual values represent the mean point on the scale '1: to a great extent', to '9: not at all'. These means indicate the extent to which respondents from each country felt that each management function (or manager responsible for each function) was involved in each stage of the buying decision.

In both British and West German samples Manufacturing and Production Engineering management play the greatest role throughout from identifying a need for a new machine tool to finally placing the order. They are the key decision takers in the machine tool purchasing decision and the main point of contact for the supplier.

Clear differences emerge in the pattern of purchasing responsibility in the West German and British companies in the sample. These differences appear to be consistent with earlier comments on the operating procedures of the companies in the samples.

One significant aspect of this process is that individual suppliers would not be involved in the decision until a need had been identified. At this stage the customer would be likely to approach potential suppliers to seek alternative solutions to the problem. In other words the customer was active in soliciting information and advice from potential suppliers. In identifying a need the customer was also specifying a technical solution to some extent. In other words the

Table 5.5. *Extent of involvement of managers at different stages in the purchasing process*

		Identifying the need	Specifying the type of machine tool	Identifying a shortlist of potential suppliers	Deciding on machine tool	Placing the order
Marketing/Sales	(WG)	7.2	8.1	7.2	7.8	6.9
	(UK)	6.1	7.5	7.8	7.9	8.3
Manufacturing	(WG)	1.5	1.2	1.9†	1.3	2.1
	(UK)	1.5	1.3	2.0	1.3	3.1
Purchasing	(WG)	5.9*	6.4	3.1†	4.8	2.8†
	(UK)	7.5	7.6	5.5	6.6	2.6
Finance	(WG)	4.1*	7.7	7.1	7.4*	5.2
	(UK)	5.0	6.5	6.8	5.5	4.5
Design Engineering	(WG)	5.7	4.7	6.0	6.1	6.2
	(UK)	5.5	4.4	6.1	6.1	7.4
Production Engineering	(WG)	2.2	1.8	2.0	1.6	2.3
	(UK)	1.5	1.6	1.3	1.2	1.2

For interpretation, see the discussion.
†Significant at the 99% level of confidence.
*Significant at the 95% level of confidence.

initial technical parameters of a solution to the manufacturing problem were likely to be determined by the customer rather than the supplier in the original definition of requirements. This 'user active' perspective clearly emphasises the importance of the quality of the customer's manufacturing and production engineering personnel.

Once initial contacts with potential suppliers have been established then the resultant specification of the machine tool required will depend on the quality of discussion between the customer and potential suppliers, and the nature of the problem. A replacement decision with little or no modification to a standard machine tool will clearly involve less discussion than the purchase of a new automatic machining system for a family of components which has not previously been manufactured.

Purchasing and Finance departments (or managers responsible for these functions) were also involved in the decision. But their contribution varied between the British and West German firms in the sample. At the stage of identifying a need for a new machine tool both functions were involved to a significantly greater extent in the West German companies in the sample. At the shortlisting stage where alternative suppliers were brought together for consideration, Purchasing had significantly greater involvement in the process in West Germany, in its traditional role of scanning the market for potential suppliers to meet a clearly defined manufacturing need.

At the decision stage the Finance function was significantly more likely to be involved in the British companies in the sample. This greater involvement had clear repercussions in terms of the importance given to financial criteria in the evaluation of alternatives, as the following sections comment. Finally Purchasing had significantly greater involvement at the final stage of placing the order for new equipment in the British companies in the sample.

These variations in the composition of the group of managers involved in the purchasing decision (buying centre) had obvious repercussions on the nature of the buying process itself. For example, the greater involvement of Purchasing in the West German companies in the sample was associated with more companies on the shortlist when German companies bought new machine tools. The greater involvement of Finance in the British companies in the sample was associated with more stringent financial criteria for new investment proposals.

Formality of investment decisions

There was considerable evidence that the machine tool purchasing process was subject to more extensive written rules and procedures in the West German companies. The overall impression of the study is one of a greater degree of planning and coordination of investment decision taking – in itself a general indicator of a higher degree of management 'quality'. (46% of the British firms compared to 20% of the West German firms had no formal rules at all.)

The degree of formality required in proposals for new investment depended on the type of investment under consideration. For most companies in the sample the replacement of an existing machine tool with one similar to the one being replaced, or the purchase of a new type of machine tool to a manufacturer's specification, required less formality than the purchase of a new type of machine tool built to the customer's own specification. There were no significant differences in the response of British and West German companies on this dimension.

It is interesting to note that the justification for purchase of purpose-built machine tools tended to be more formal in the majority of companies in the sample. As the technology becomes more complex and the uncertainties grow the customer has to make a more sophisticated case for the investment. If the customer does not have the necessary expertise to make such a case effectively then the more complex 'custom-built' project may be more difficult to justify. The data support the importance of customer sophistication in making the case for investment where a greater degree of product modification is required.

The discussion of proposals was significantly more formal in West German companies in the sample as Table 5.6 indicates. Informal discussion appears to be a more significant aspect of British investment decision taking, combined with formal written proposals. By contrast the West German approach appears to be significantly more formal, associated with an investment planning meeting, at which new proposals are discussed.

The greater degree of formality and control apparent in the investment decisions of the West German companies in the sample appear to the writer to be consistent with the general theme of management quality developing in this review. The more extensive use of formal investment planning committees, extensive written rules

Table 5.6. *Formality of discussion of proposals for purchase of a machine tool*

Question: If more than one manager is usually involved in the decision to invest in new machine tools, how are proposals for investment discussed?
(Please tick where appropriate – you may tick more than one category where necessary.)

Method of discussion	West Germany	United Kingdom
Informally as proposals are put forward	30*	58
Formal proposals are prepared and distributed	26	55
To an investment planning meeting	50	13
Base	100 (56)	100 (73)

*To be read: 30% of the respondents from West German companies indicated that one way of discussing investment proposals in their company was informally as proposals were put forward.
Each of the differences between the British and West German companies was significant at the 99% level of confidence.

and procedures, and a greater degree of formality in the preparation of proposals, all suggest a more careful systematic approach to machine tool purchasing.

Machine tool purchasing decision

The techniques used in the machine tool purchasing process are central to the study of customer behaviour since the final decision on a specific proposal will reflect the way in which the proposal has been formulated. This part of the study therefore is concerned with an in-depth comparison of the techniques used by West German and British companies in machine tool investment decisions. It takes the perspective of an active customer which defines a problem, seeks alternative technical solutions, then takes a final decision between these alternatives.

During the course of the field research it became apparent that decisions to buy new machine tools were precipitated by two major factors. Either the costs of operating existing equipment had passed a defined threshold, or changes had occurred in the demand for certain types of machine (caused in the majority of cases by the introduction of new end-products, an expansion in the demand for existing end-products or the redesign of a major component). The principal cost changes influencing the decision to invest in new equipment generally were increases in labour costs. However, there was a constant pressure on companies in both countries towards saving skilled labour, as a relevant factor in all investment decisions – this would be unlikely in itself to precipitate the replacement of a specific piece of plant.

Increases in the costs of specific plant through unplanned downtime or increased unreliability of output were more likely to provoke change as managers sought new and more efficient plant. The need to catch up with best practice techniques was clearly also apparent. The importance of this factor clearly depends on the extent to which individual companies collect and use information on the costs of existing plant. 61% of the total sample collected information on the costs of operating individual machine tools, usually on a monthly basis. 50% of the sample stored the data on card index systems, whilst 36% used computer-based systems.

West German companies were significantly more likely to collect and assess the operating costs of individual machine tools on a regular basis (76% of the West German companies compared to 50% of the British companies). By doing so they were more likely to identify unacceptable changes in the operating costs of existing machine tools through increased unreliability, and the extent of differences between existing costs and the costs of best practice techniques (incorporating technical change). The cost-based stimulus for change appeared to be more likely to be identified in the West German companies in the sample.

Change in demand for the output of the machine tool was the other significant precipitating factor. Again there was a significant difference between the West German and British companies. The West German companies in the sample were significantly more likely to forecast the utilisation of individual machine tools (91% of the West German companies compared to 63% of the British companies in the sample). This is another indicator of a more care-

Table 5.7. *Methods of setting budget for capital expenditure*

Method	West Germany	United Kingdom
By reference to what the company spent last year	31%*	5%
By reference to what the competition are spending	26%	2%
By reference to what the company can afford	36%	74%
On the basis of specific estimates of production requirements	87%	87%
Base	56	73

*To be read: 31% of the West German respondents indicated that their capital expenditure budget took into account what the company had spent last year.

ful, planned approach to investment decisions based on projections of future demand for equipment.

Once the need for equipment has been identified for most companies in the sample this became part of a budget for capital investment for the following financial year. Competing investment proposals from production management were evaluated and combined in an overall investment budget according to priorities agreed by management as a whole. The greater formality of the West German approach has already been commented upon.

There were some significant differences in the way in which the annual capital budget was set as Table 5.7 indicates. The West German approach stressed a greater degree of long term planning with greater reference to expenditure in the previous financial year, and a greater consciousness of competitive activity. By contrast the British respondent was more concerned with what the company could afford, once more stressing the role of the accountant and the importance of financial considerations.

Once a decision had been taken to go ahead with the purchase of a machine tool, it was common procedure to draw up a shortlist of potential suppliers. 80% of the companies in the sample did so for a replacement of an existing machine with one of a similar type (replacement buy), 89% for a new machine tool built to the manufacturer's specification (supplier-specified) and 76% for a special purpose machine tool built to the firm's own specification.

There were no significant differences between the number of firms on the shortlist in each country for straight replacement or customer-specified items. In both countries most companies had three suppliers on the shortlist. There was a slight difference between the number of firms on the shortlist for supplier-specified machine tools which the user had not bought before (for example, a new generic type of machine tool such as a CNC (computer numerical control) lathe). The West German companies in the sample were likely to include more suppliers on their shortlist than the British companies (four suppliers on average compared to three). There was a tendency therefore to look more widely and in considering more alternatives to consider new types of machine tool. The number of companies shortlisted was relatively small for all categories of investment for the sample as a whole. This was one measure of a relatively narrow pattern of search in each country indicating a considerable degree of source loyalty or inertia – assuming that the same supply companies were consulted on each occasion.

The criteria used by respondents in evaluating the general suitability of a supplier before shortlisting are listed in Table 5.8 together with the ratings of importance which respondents gave to each criterion. From this table it is clear that the most important factors were the technical excellence of the supplier and aftersales service and spares. Ease of communication with the supplier and past experience with the company were the next two factors of importance followed by general business and financial reputation. Country of origin was the least important factor. These ratings are consistent with the results of the study of machine tool buyers conducted by NEDO for the machine tools EDC (1965). Technical excellence is the most important factor given here by a sample of British and West German buyers when assessing the suitability of a supplier to meet their needs (again stressing the importance of product quality as a factor in competition). The quality of aftersales service and spares is also clearly crucial, again consistent with the results of the NEDO survey.

This research is ultimately focussed on the relationship between a supplier and its customers in new product development, and it is

Table 5.8. *Rating of the importance of general supplier attributes*

Question: How important are the following general factors when you assess a supplier's suitability?

Supplier attribute	Rating: Very important 1+3	5	Not important 7+9	$\bar{x}$	*n*
Domestic rather than overseas based	41*	31	27	4.6	121
Financial position	58	27	15	3.8	114
Technical excellence	97	2	1	1.3	123
Quality of management	50	35	14	4.2	112
Ease of communication with supplier	73	22	4	2.8	115
Aftersales service and spares	96	3	1	1.5	124
General business reputation	79	12	8	3.1	121
Past experience with company	77	17	6	2.8	118

*To be read: 41% of the total sample indicated that the fact that a potential supplier was domestic rather than overseas based was an important factor rating either 1 or 3 on a scale from '1: very important' to '9: not important'.

$\bar{x}$ = mean rating assuming the scales have interval properties.

n = number in sample answering this question.

significant that ease of communication with the supplier and past experience with the supplier are ranked as the next most important factors. This emphasises the general importance of the relationship between the firm and its customers. In addition the relatively high ranking of 'general business reputation' indicates that the relationship between a supplier and other customers is also important to the individual company when evaluating a potential supplier's suitability. The rating of country of origin is less important than the other factors for most respondents, although for a substantial minority (41%) it was important.

Technical excellence was the only factor which was rated significantly differently by respondents from each country. While the distribution of response was concentrated on the rating 1 or 3 there was still some variation between the two samples. 96% of the West German respondents rated technical excellence at the first point 'very important' on the scale compared to 82% of the British respondents. The overall rating of technical excellence as an important factor in the evaluation of the suitability of a supplier was therefore slightly higher amongst the West German companies in the sample.

The sales engineer was seen as the most useful source of information by the respondents in the machine tool purchasing process followed by exhibitions, members of the buyer's own company, managers in other firms and the supplier's service engineer. Advertisements in journals and leaflets and brochures were seen as less useful. Given the importance of the machine tool salesman it is clearly useful to examine which factors appear to influence his performance as far as the buyer is concerned. Table 5.9 illustrates the views of the sample as a whole.

Again technical considerations predominate. Technical training and product knowledge are ranked as the most important factors influencing the salesman's performance, followed by commitment and motivation, knowledge of market trends and creativity. Languages were also seen as important, but financial factors such as high basic salary and incentive schemes were seen as significantly less important.

Three of these attributes were rated significantly differently by respondents as Table 5.10 shows. 'Commitment' and 'status' were seen as significantly more important by respondents from the United Kingdom than West Germany, and 'creativity' was seen as significantly more important by respondents from West Germany ('creativity' was defined as the ability to find new technical solutions to manufacturing problems).

Differences between British and West German companies on the rating of 'creativity' are large and have a major consequence. If the West German user was more likely to rate creativity as an important factor then he may be more receptive to new solutions to a problem than the British user, i.e. more receptive to technological inno-

Table 5.9. *Importance of salesman attributes*

Question: How important would you say that the following factors were in influencing the performance of a machine tool salesman?

	Rating				
Salesman attributes	Very important 1 + 3	5	Not important 7 + 9	$\bar{x}$	n
Knowledge of market trends	74*	20	6	3.0	115
Technical training	94	4	2	1.8	123
Product knowledge	84	10	6	1.4	114
Languages	65	24	10	3.5	116
Commitment and motivation	82	16	2	2.4	115
Incentive schemes	24	33	27	5.4	113
Creativity	72	18	11	2.9	120
High basic salary	39	35	26	4.7	112
Recognised high status	41	33	25	4.7	115

*To be read: 74% of the respondents indicated that the knowledge of market trends was an important factor in influencing the salesman's performance rating either 1 or 3 on a scale from '1: very important' to '9: not important'.

$\bar{x}$ = mean rating assuming the scales have interval properties.

n = number in sample answering this question.

vation. By contrast the British user appears to be preoccupied with more traditional values of 'status' and 'commitment' – encouraging the domestic supplier to give greater emphasis to this aspect of salesman performance.

At the investment appraisal stage the West German respondents appeared to be more concerned with the technical characteristics of alternative competing proposals, whilst the British customer appeared to be more concerned with economic factors, principally price and operating costs. It was felt that this concern might be derived from the greater involvement of Finance at the decision

Table 5.10. *Differences in the rating of commitment, status and creativity (British ratings in brackets)*

	Rating				
	Very important 1	3	5	7	Not important 9
Commitment and motivation	35* (57)	39 (32)	26 (11)	—	—
Status	5 (15)	15 (42)	37 (30)	19 (6)	23 (7)
Creativity	82 (18)	13 (35)	6 (27)	0 (20)	—

*To be read: 35% of the West German respondents replying to this question rated commitment and motivation as a very important influence on the performance of a salesman scoring 1 on a scale from '1: very important' to '9: not important'.

stage in the purchase of machine tools in the United Kingdom (see Table 5.5). Appraisal techniques reflected these differences. British companies in the sample were significantly more likely to include estimates of operating and waste costs in their calculation. They were also more likely to include estimates of production utilisation and programming costs.

Close attention to this type of information can inhibit the rapid acceptance of technological innovation, particularly when the benefits and costs of adopting the innovation are not clear. Where there are uncertainties in these cost areas the manager responsible for evaluation may well prefer a conventional solution with more predictable costs. The decision to invest in new technologies does not lend itself to conventional investment appraisal techniques involving detailed cost estimation. This may well be one of the most important factors inhibiting the rapid response of user companies in the United Kingdom to new technologies. This point is developed further in the discussion of the final chapter.

This can be seen in the use of pay-back techniques. A greater proportion of British than West German companies used some form of pay-back analysis (66% compared to 55%). (DCF (discounted cash flow) techniques were used in 39% of the British companies compared to 21% of the West German companies.) There was also

Table 5.11. *Investment pay-back periods UK and West Germany compared*

Period	West Germany	United Kingdom
Up to 30 months	14%	25%
30–42 months	19%	40%
42–48 months	22%	11%
More than 48 months	46%	23%
Total	100% (53)	100% (76)

Percentages may not total 100 because of rounding.

a significant difference in length of time a new investment was required to pay back its costs as Table 5.11 shows.

The pay-back period for many British companies in the sample was clearly shorter than that of the West German ones. For example, 68% of the West German sample set a pay-back period of longer than three and a half years compared to 34% of the British sample.

Such practices will inhibit the early acceptance of innovation in the British sample. If an investment is required to pay back its initial cost rapidly then this favours the conventional technology with predictable performance. There is no time for potential 'teething troubles' to be ironed out of a new technology, and although the potential returns from adopting the innovation may be higher, these are likely to be discounted against the uncertainties and potentially higher costs of assimilating a new production technology into the existing operation.

Concluding remarks

The survey of West German and British machine tool users reported in this chapter has provided some support for the 'user active' viewpoint which suggests that the problems of the British machine tool industry may be caused, in part at least, by the domestic customer. The chapter has provided several indicators that technical factors are more important to the West German companies in the sample. This was clear not only in the importance rating given to the attributes of the machine tool, but also in the general rating of the importance of the technical competence of the supplier and the machine tool salesman. The West German customer, it appears, it more likely to expect and demand technical qualities from its suppliers.

There are also several indicators of differences in the general quality of management which appear to have an impact on investment decisions. The West German company tended to be relatively more formally controlled, with shorter reporting cycles for management, and longer planning horizons. At the same time, since many of the West German companies in the sample were independent companies rather than members of a larger group, performance targets could be set in other than strictly financial terms. By contrast, in the British companies in the sample the performance target was frequently financial, with the accountant dominating decisions. These conditions are not conducive to the rapid acceptance of new technologies with pay-back periods which by definition are uncertain.

The inhibiting role of the accountant is also clear in the machine tool purchasing decision itself. In the British company he was more likely to participate in the final decision process and the result of this participation was evident in terms of setting budgets in the light of what the company could afford rather than in terms of what the competition was doing, and in setting far shorter pay-back periods for new investment. Conventional investment appraisal techniques, more extensively used in the British companies in the sample, do not favour new technologies with uncertain pay-back periods.

There were also further indicators of general differences in the quality of investment decisions. These included greater attention to the preparation of a formal case for investment in the West German companies in the sample, and more extensive formal consideration of the decision.

At the same time in making a case for new investment it is clear that the customer is developing a solution to a manufacturing problem with a specific supplier. The importance of such customer–supplier interaction was confirmed in an examination of a series of general factors where past experience and ease of communication were clearly rated as important.

6

Buyer–seller interaction in new product development

Introduction

In Chapter 6 emphasis changes from the activities of individual companies to the relationships which have developed between buyer and seller. Whilst Chapter 6 attempts to isolate the interaction between buyer and seller as the focal point for discussion, it is clear from the discussion in Chapters 4 and 5 that both 'supplier' and 'buyer' interpretations already have indicated the importance of this interpretation. Discussions of these differences will not be repeated in detail in this chapter. Rather, attention will focus on differences in the attitudes to interaction of British and West German machine tool manufacturers and their customers. These differences appear fundamental and help to explain some of the problems encountered by this industry in the United Kingdom in developing new products in line with the market's needs.

Supplier attitudes to involving customers in product development

In Chapter 4 considerable differences were identified in the extent to which West German and British machine tool suppliers involved the customer in product development work. At the idea generation stage there was greater contact between the design engineer in the West German company and potential customers. Design managers spent longer in the field than their British counterparts and were involved to a greater extent in day-to-day contacts with the companies' customers. The West German supplier also tended to involve the customer to a greater extent in establishing the initial feasibility of the new product idea in West Germany, and the idea was more likely to be taken to two or three chosen companies to establish the market's reaction. The customer was also more likely to be involved in the evaluation of prototypes in West Germany, and the prototype would often be tested in a customer's works over an extended period. In the United Kingdom the prototype (if one was built) was more likely to be built and tested in house by the manufacturer. Effective new product development was associated with greater and more continuous involvement between the supplier and potential customer.

Differences in techniques of new product development clearly reflect the motivation and experience of the managers involved. They also reflect the attitudes of individual managers to the value of involving the customer. During the interviews specific comments were made by individual respondents about the extent to which the customer tended to become involved. Table 6.1 contains extracts from these comments which serve to illustrate the sample's views of the importance of interaction with customers.

The British managers making these comments were generally less positive about involving the customer in product development work than the West Germans. Three main factors appear to be conditioning this response.

The first was the view that a degree of secrecy was necessary to develop and launch a new product successfully in the market. Companies expressing this viewpoint felt that they could not defend themselves for long against competition if they developed similar products. It was necessary to keep the development secret until market launch and then seek to obtain rapid market penetration before the competition could react. But this would be a somewhat futile exercise if the new product failed through lack of attention to customer needs. The need for secrecy was also supported by the suggestion that if a potential customer became aware of a new product being developed then they might postpone orders until the new machine was on the market, and the manufacturer would lose orders meantime.

From these comments it also appears that some British managers felt that their domestic customers lacked the necessary technical sophistication to evaluate ideas for new machine tools, and contribute to their development. Therefore, it was not useful to attempt to involve the customer until the market launch, and it was the supplier's responsibility to develop new products by identifying

Table 6.1. *Comments on interaction with customers (the supplier's view)*

West German suppliers
Company C 'If we have an idea on how to develop a new product, which has come from Technical Development or a customer, then the first thing we do is to approach two or three customers and ask them what they think about the idea, and if they have any changes to suggest.' 'We know perhaps 80% of these conditions (customers' needs) very accurately, but to get the remaining 20% we have to talk carefully with our customers about their needs.' 'I think it's impossible to develop a machine without very close contact with the customers . . . occasionally we will develop something secretly to avoid competitive copying but it's usually very important to give news on new products to our customers as quickly as possible, especially if they are at the planning stage of a new production development. You need to be involved in their production planning and development projects as early and as fully as possible.'
Company D 'We usually try to get a customer who is close at hand, not necessarily technically qualified, and capable of not saying anything about the development until it is proven. We try to get a customer who is a good friend and close to us, where we can do a quick service and check the machine daily where necessary. We should have a good relationship with them.'
Company E 'You must solve the customer's problems more and more, and as cheaply as possible. It takes a long time to get the design people to think marketing. Ten years ago we used to make a lot of standard machines, now we make more specialised machines and the initiative comes from the customers.' 'We always change who we go to with new machines – in some instances it is difficult to get good relationships with particular companies.'
Company F 'As far as getting deliberately involved with the customers – I think that is something automatic. As a relatively old company we have many long established customers all over the world which we are in contact with.' 'It is 99% the reliability of the product which creates good relationships – quality in terms of how efficiently the machine works, and then reliability.'

Table 6.1 (*cont.*)

Company G
'If there is agreement in general terms about a product idea then the Sales Department will approach customers to ask them about the general concept. These customers are managers with whom we are in close and continuous contact . . . they want innovation because it is in their own interests to be seen to be using leading techniques of production.'

Company H
'At the final stage of development the company again goes to the market to make sure that the product meets the market's requirements. At this stage the aim will be to find user companies where the potential buyer is familiar with the problems of machine tool construction, possibly through a formal training in the industry . . . '

British companies

Company J
(Speaking of customer involvement at design stage)

'The design is not taken to customers at that stage – if they get the wind of a new product in the offing, then they will wait for the new product and you will lose orders. We don't do a direct questionnaire survey of customers' needs because the customers will ask for all sorts of different things – usually with confusing results.'

(Speaking of customer involvement at prototype stage)

'Customers don't want to have to solve all the problems in a machine, and to do this [extended trial of the machine in the works], you need a customer who is sympathetic with your company, and understands your problems . . . you need someone who is prepared to stick his neck out, and there are not too many of these in British manufacturing industry – certainly not amongst our customers.'

Company K
'I don't think customers have enough farsighted vision to specify new machines in isolation, to be in constant communication irrespective of demand. We get to hear about it when they have production needs – possibly when they have a new product of their own and they want to increase output . . . But we don't exist to solve the production engineering problems of Lucas or Rolls Royce for example – we exist to sell as many machines on as broad a front as we possibly can – machines need to be evaluated against the overall market demand – we must develop products which stand a chance of selling on the widest possible front and not one particular to one market.'

Table 6.1 (*cont.*)

Company L
'In the United Kingdom one of the biggest problems is the reluctance of customers to get involved in the development of new machines – there is an inherent conservatism of the customer, and a reluctance to accept the responsibility of being first . . . the customer is rarely involved in the design of "standards" – he wouldn't usually be involved until the machine was shown with dramatic effect at its first exhibition.'

'There is a great difficulty in introducing high technology in the UK, because of the buyer's reluctance to be first. It was never Joe who bought the first machine, it was Fritz.'

Company M
'We certainly have major customer types who buy a lot of our machines, but they will not necessarily be involved in anything new that we are doing – we would tell them about it and look on their response with interest . . . they may have some influence if they come along and say why are you doing that . . . we may rethink if we value the company's judgement, and the person's judgement . . . '

Company N
(Speaking of customer involvement in design stage)

'It tends to be minimal because it is very rare that the customer is going to clearly originate the new concept – for example, if it were left to customers we still would not have CNC machines – this was a technology lead advance that was the answer to a production problem. However, there are some large companies who it is worth listening to – they tend to be the companies with specialists whose job it is to think ahead and decide what their production requirements are going to be.'

Company O
(One of the few exceptions to the general British response)

'We have a very small world market for our core product – it's very specialised and we are very close to our customers. Within our sales force there is a lot of product knowledge, and they know the people who would use thread grinders very well. The engineering function is close to the customer because there is quite a lot of engineering application and close direct liaison.'

'One of our customers is Daimler Benz, to whom we have sold a machine for grinding ball screws and steering worms. The machine is a better product because we sell it to them.'

market needs (the customer playing a passive role). The final factor was the view that British customers were inherently conservative and unwilling to accept new technologies until they were well proven. Consequently they were unwilling to become involved with the supplier in pre-launch activities such as prototype development and testing.

Customer attitudes to involvement in new product development

From the discussion of Chapter 5 there is evidence that British customers were less receptive to new technology than their West German counterparts, and less likely to demand it from their suppliers. This was reflected in the extent to which they saw themselves as involved with the supplier in the product development process. This involvement can be measured in a variety of ways, for example by a count of the number of supplier visits to specific companies and the level and extent of contact between different management functions over time. It can also be measured in terms of the number of joint programmes between supplier and customer when new technologies have been required to solve specific customer requirements.

However, in this study only one simple measure of involvement in product development was developed in the study of the customers. This measure asked the machine tool buyers in each sample to indicate the extent to which they as customers became involved in three types of product development work: continuous minor developments in the machine tool itself (small scale incremental innovation), the development of a completely new type of machine tool to a supplier's specification, and the development of special purpose machine tools built to the company's own specification.

The relative importance of each of these categories of investment was indicated by the response to a question asking respondents to indicate the proportions of buying decisions (by number of machine tools) which fell into each category. There were no significant differences in the response from either country. The average proportions for each category were 39% replacement buy with little or no modification, 35% new type of machine tool with little or no modification to the supplier's specification, and 10% special purpose machine tools to the customer's own specification. These figures illustrate the relative importance of supplier-specified rather than buyer-specified innovation. They also suggest that the replacement of an existing machine tool with one which was basically similar was the most frequently occurring type of buying decision.

Table 6.2 provides a measure of the extent to which customers felt that they became involved in the development of new types of machine tool. The response to this question suggests that there was considerable variation in customer involvement in both countries for all three categories of change.

Customer involvement is clearly greatest for the development of special purpose machine tools built to the company's own specification, as would be expected, although a considerable proportion of respondents clearly felt that they played no role in suggesting a technical solution to their manufacturing problem and modifying the product idea as it developed. It is also evident that customer involvement was likely to be greatest at the initial stages of idea suggestion rather than at the stages of evaluating and modifying the development as the design is developed and a prototype built. There is evidence therefore to support both supplier-active and customer-active views of product development. It appears that both supplier and customer could take an active role in the product development process.

Comparing the responses of British and West German customers there was some evidence that West German customers played a greater role in suggesting the original idea for change than the British customer. Table 6.3 contrasts this involvement.

The West German customer appears to play a more active (albeit limited) role in suggesting ideas for minor product change and new types of machine tools than the British customer. The results of the comparison between customers in the previous chapter suggest how the general quality and orientation of management in the West German companies would tend to encourage this difference.

Summarising the evidence on buyer–seller interaction in this study from the manufacturer's perspective, West German companies saw interaction with their customers as an essential part of the product development process, and sought to encourage customer involvement at the early idea generation and design stages, and at subsequent stages of prototype development and testing. In contrast British companies were less active in seeking customer cooperation believing that this was not helpful in product development. This negative attitude stemmed from a view that their own

Table 6.2. *Customer perception of involvement in product development*

Questions
(a) To what extent does your company as the buying company become involved with the supplier in the continuous minor development of machine tools?
(b) To what extent does your company as the buying company become involved with the supplier in the development of completely new machine tools to the supplier's specification?
(c) To what extent does your company as the buying company become involved with the supplier in the development of special purpose machine tools built to your specification?

	To a great extent			To some extent			Very little			Not at all					
	(a)	(b)	(c)	(a)	(b)	(c)	(a)	(b)	(c)	(a)	(b)	(c)	(a)	(b)	(c)
In suggesting the original idea	9%*	7%	26%	14%	11%	18%	26%	23%	13%	51%	58%	43%	123	122	117
In evaluating and modifying the development as the design is developed	8%	2%	17%	16%	13%	23%	22%	24%	14%	54%	60%	46%	123	120	114
In evaluating and modifying any prototype	4%	1%	15%	17%	16%	21%	25%	21%	16%	54%	62%	47%	121	118	112

*To be read: 9% of the respondents (123 in total) indicated that they were involved to a great extent with the supplier in suggesting original ideas for the continuous minor development of machine tools.

Table 6.3. *Customer involvement in product development*

	To a great extent	To some extent	Very little	Not at all	Total
Extent customer perceives involvement in suggesting idea for a minor product change					
West Germany	17*	11	23	49	100 (53)
United Kingdom	3	16	29	53	100 (70)
Extent customer perceives involvement in suggesting idea for new type of machine tool					
West Germany	25†	—	27	47	100 (51)
United Kingdom	14	20	—	66	100 (71)

*To be read: 17% of the West German customers in the sample said that they felt they were involved to a great extent in suggesting the idea for a minor product change.

†To be read: 25% of the West German customers in the sample said that they felt they were involved to a great extent in suggesting the idea for a new type of machine tool.

Percentages may not total 100 because of rounding.

domestic customers were not sophisticated enough to assist them in the development of new technologies, and would be unreceptive to such technologies once developed. West German suppliers also appeared to have developed long term relationships with a few selected customers to whom they referred new product development projects. Such relationships continued independently of individual product development decisions. From a customer's perspective the West German companies appeared to be more likely to become involved than British companies in the supplier's development activities, particularly in suggesting the idea for product modification and new types of machine tools. In short the West German customer played a more active role in new product development than the British customer.

The comparative isolation of British manufacturers from their customers could therefore explain many of the problems of developing new products in line with the market's requirements.

Factors influencing the extent of customer–supplier interaction

Given the evidence of preceding sections which suggests how closer buyer–seller relationships appear to be important in effective new product development, it is relevant to examine the factors which may influence this relationship. An understanding of these factors may then help to provide guidelines for an organisation wanting to improve its relationship with its customers (or suppliers).

One of the major determining factors in the extent to which buyers and sellers develop a continuing relationship is likely to be the degree of concentration in the market itself. Company O was one of the few successful exporters in the British sample of manufacturing companies (exporting more than 70% of its output in 1978). It had a very small market for its core product, a specialised thread grinding and tapping machine, and consequently was in close and continuing contact with its customers. There were also relatively few competitive suppliers for the customer to choose from.

By comparison company N's products were basically standard items which were sold to a wide variety of end users, and the degree of involvement with individual customers was minimal. In general as the number of suppliers and/or number of customers increase, so buyer–seller relationships might be assumed to become less permanent, since each would have more potential partners at each new transaction.

However, this ignores the effect of managerial factors such as the extent of search for sources of supply. The customers in the survey did not look far in their search for new alternatives (the average number of firms on the shortlist was three), which meant that even if the number of potential suppliers was high the individual buyers were only considering a limited subset of these sources. Thus, there is apparently a phenomenon of inertia or source loyalty which may lead to constrained search even if the number of potential suppliers is relatively large. The costs of changing partners from one purchase to the next are clearly an important limiting factor, and would reinforce existing buyer–seller relationships. Such costs include the time needed to assess the general capabilities of a new supplier, and the 'learning costs' associated with adopting a new supplier such as procedures for invoicing, spares handling or servicing.

Absolute measures of market concentration in terms of the number of buyers or sellers in a particular market segment are therefore of less use in interpreting the patterns and strengths of existing relationships, than the extent to which buyers and sellers perceive the market to be concentrated. There was no evidence that

British and West German companies differed in the latter respect in this enquiry.

Previous work in the administrative sciences has suggested that size of firm would have an important impact on the extent of inter-organisational relationships. As organisations decreased in size it has been argued that interdependence would increase as organisations combined resources to accomplish a given task. This concept was extended to predict buyer–seller interaction, and it was suggested at the beginning of this study that there would be greater interaction between smaller companies than larger ones.

However, the results of the survey of customers suggested that buyer–seller interaction could be expected to increase rather than decrease with the size of company. Large customer-companies were significantly more likely to become involved by suggesting the original ideas for minor developments, and new types of machine tool. Once again size of company appears to be acting as a surrogate measure of other factors such as the range of technical skills available to the buying company. As the company increases in size so it is more likely to employ specialist managers capable of discussing new technologies with the supplier. Its purchasing power can also be expected to increase, and with this its influence on supplier development work.

The type of contact also appeared to change with increase in size. For example, in Supply Company B it was felt that in small companies many of the customers were 'toolroom men, they are expert in their own field, they know what they are saying and they know exactly what they want'. By contrast in large companies 'the Financial Department gets more influence and emphasis changes from technical points to financial considerations'.

There were significant differences in the degree of centralisation of companies in the British and West German companies in this study. Given a higher degree of user–supplier interaction in the West German companies and a greater degree of decentralisation in control, there was some support for a relationship between the degree of centralisation of a company, and the interaction between buyer and seller. As organisations (buying or selling) had more discretion over their individual activities, so buyer–seller interaction increased.

Historically the British machine tool industry has become increasingly centralised, with individual companies being acquired by larger financial groups, often with holdings in other engineering interests. As the industry note (Appendix A) comments this may well inhibit individual supplier–customer relationships. Whilst concentration and large-scale manufacturing may reduce the unit costs of standard products, it is not in itself conducive to the production of special purpose technically sophisticated products built to customers' individual needs, and it would appear that this (special purpose products) is more in line with the needs of the market.

There were some signs that the practical implications of this were being recognised by British suppliers at the time of the survey. For example, in companies L and M (British), group management policy was now to give back as much autonomy as possible to the individual member companies, particularly in the areas of design and development. In this way it was hoped that individual companies could become more responsive to individual customer needs.

There was considerable support for the view that customer–supplier interaction would be greatest when both had many of the characteristics of the technically progressive firm (identified in Chapter 2). From the analysis of Chapters 4 and 5 it was evident that customers and suppliers in West Germany generally exhibited more of these characteristics than did their British counterparts. For example, West German customers were significantly more likely to belong to and participate in professional and technical associations than British customers and gave greater attention to technical requirements rather than financial criteria in decision taking. The engineer was significantly more likely to be involved at higher levels in the direction and control of West German companies (both supplier and customer), and technical excellence was stressed as an important criterion in supplier choice more by West German customers than by British customers.

There was no support for the view that customer–supplier interaction would be greatest when customers and suppliers had similar organisational climates. Generally the West German companies (customer and supplier) tended towards the mechanistic model, whereas the British companies (customer and supplier) tended more towards the organic model. However, interaction was greatest between the West German customers and suppliers which had similar mechanistic characteristics, and least between British customers and suppliers which had similar organic characteristics. (If anything, greater interaction might have been predicted between organic rather than mechanistic customer and supplier organisations.)

One underlying cause of this is potential difficulties of dealing with an organic organisation (where responsibilities are not clearly laid down and identifiable, and where tasks are shared between individual managers). These conditions may increase the initial receptivity of a customer organisation to innovation or create a better environment for new product development in the supplier organisation, but will not in themselves facilitate easy communication between buyer and seller during the development process itself.

There was considerable support for the view that buyer–seller interaction would be highest where formal procedures were used in the new product development and product purchasing decision. West German suppliers were significantly more likely to use formal techniques to develop new machine tools, and their customers used correspondingly formal techniques in evaluation. A higher degree of interaction was present therefore between companies which gave equal and high significance to activities such as value engineering in design and product cost analysis, a pattern which was more likely in West Germany than in the United Kingdom.

Summary

The analysis in Chapter 6 confirmed the view that the degree of interaction between customers and suppliers in West Germany was greater than in the United Kingdom in the industry studied. However, the relationship between customer–supplier interaction and successful new product development was seen to be tautological since successful new product development involved greater customer–supplier interaction. This had already been shown in the analysis of product development decisions in Chapter 4. Nevertheless the analysis in Chapter 6 supported the overall conclusion that British companies were less successful in product development in certain industries because the degree of interaction between themselves and their customers was limited.

7

Implications of the study

Introduction

The research reported in this book was prompted by the observation that British companies in certain sectors of the mechanical engineering industry were less successful in developing products of the right quality to meet the needs of the market than some of their principal overseas competitors. This was seen as a major factor in explaining the relative decline of these sectors in terms of international market share, and a high rate of growth in import penetration into British markets. The machine tool industry was selected for study and comparisons were made between the new product development activities of British and West German machine tool manufacturers. Comparisons were also made between the attitudes of customers in each country to new technology, and their approach to purchasing new types of machine tool. This final chapter examines the implications of this study for machine tool manufacturers and their customers. It also considers the implications of the study for policy makers concerned with improving the performance of this industry and other related industries.

Manufacturers of machine tools

Comparisons between the British and West German machine tool manufacturers identified several key differences in their approach to new product development. West German companies had greater direct involvement with their customers in the design and development of new machine tools. Their design staff spent more time in direct contact with customers and interpreting sales information than their British counterparts. They spent longer periods testing and evaluating prototypes and were more likely to test the prototype in a customer's plant under realistic working conditions. West German companies were also more likely to design special purpose machine tools to individual customer specifications and give greater importance to technical quality in product design.

The importance given to such technical factors was evident in the greater attention which was given to engineering skills rather than financial expertise as a qualification for general management. West German companies were more likely to be independent companies, owned and managed by descendants of the original founder and maintaining engineering traditions. Such companies were able to justify their performance in technical as well as strictly commercial terms.

Contacts with universities and other research institutions were also more extensively developed than in the United Kingdom, strengthening the quality of the companies' research and development activity. West German academics also appeared to give equal emphasis to applied and theoretical problems when working with machine tool manufacturers. In this sense no clear distinction was made between applied and theoretical problems.

The research indicates that the British machine tool manufacturers in the sample should make some fundamental changes. Specifically, they should increase the extent and quality of their involvement with domestic customers at all stages in product development work from origination of the idea through the evaluation of alternative designs, to the testing and evaluation of prototypes. To achieve this more attention must be given to the substance of marketing: that is to ensure direct contact between customers and product design, to strive continuously to modify or develop products in line with customer needs (stressing product quality as well as price) and to ensure that this is achieved at a profit to the company through close attention to cost control. The introduction of formal marketing departments with associated formal market research programmes may facilitate this process, but should be regarded as a trapping rather than the substance of the process.

Machine tool customers

Successive reports on the British machine tool industry (and, indeed, on the British mechanical industry as a whole) have stressed the problems of suppliers in developing new products to meet the demands of the market. However, the findings of this study suggest that these approaches have been myopic because they have tended to ignore the contribution which the customer organisation makes to product development activities. Given the importance of user contributions apparent from this research, it is clear that recommendations to improve the performance of British industry in these sectors need to take into account the weaknesses of the domestic customer, as well as the domestic supplier.

In order to improve the performance of the machine tool industry as a whole this research has indicated that several changes would be needed in British customer organisations. Specifically, the relative status and recognition of Engineering and Purchasing in the organisation should be improved. There are already signs that both these functions are being given increasing recognition, and this trend should be encouraged at every possible level by government and professional institutions. British customers need better education in the use of systematic procedures to monitor the costs of operating existing machine tools and in predicting the level of utilisation of existing machine tools. British customers also need to be encouraged to take a longer-term view of investment decisions.

Companies should be encouraged to think in terms of the quality of output from a new machine tool and the impact of changes in technology on a whole range of associated costs. For example, traditional financial evaluation procedures when applied to new technologies often fail to take account of the additional benefits of the new technology, and strict application of conventional appraisal criteria would lead to their rejection. Finally, the formality of the machine tool purchasing process should be increased with more formal documentation required to justify purchasing recommendations, and more systematic procedures developed to identify alternative solutions to the manufacturing problem, and to evaluate the solutions generated.

Customer–manufacturer relationships

The results of the study suggest that the interaction model could be used to improve the performance of an individual company in new product development in technically complex areas. Customer contact does not begin at the commercialisation stage in the product development cycle as much of the traditional literature discussed in earlier chapters assumes it does. In reality this contact persists independently of individual product development decisions. The successful supplier is likely to cultivate and develop customer organisations (referred to as 'friends' by one respondent) to help with the process of new product development from the original point of idea generation, through evaluation of alternative design ideas, the development and testing of prototypes and ultimately in market development through early adoption.

This research has suggested that such customer organisations will tend to be relatively larger than average in size, and more decentralised than other customers. They will also have many of the characteristics of the technically progressive firm, but paradoxically perhaps will be more mechanistic or formal in their general assignment of responsibilities and in the formality with which they discuss investment proposals. The latter formal characteristic may inhibit rapid internal acceptance of innovation, but will facilitate a greater degree of contact between organisations where managers in each can quickly identify points of communication and decision criteria.

In order to improve customer relationships it is likely that manufacturers will have to develop many of these characteristics themselves. The research suggests a form of organisational 'matching' where customer–manufacturer interaction was greatest between organisations with the general characteristics discussed above (i.e. large size, decentralisation, formal procedures used in decision taking and formal recognition of responsibilities, and technical progressiveness).

From the manufacturer's point of view it is clearly important to create the impression in the customer's mind that the manufacturer has similar attributes to his own organisation. In new product development work this research has suggested that the customer will value the general characteristics identified above. If the manufacturer does not match up to these characteristics then some degree of 'matching' is required to change the manufacturer to meet the customer's perception of the type of organisation which would be most suitable to work with in new product development. In some instances this may require a fundamental restructuring of the manufacturer in order to change its appearance to the customer and encourage the necessary degree of interaction. New product development decisions then become a problem of organisational design.

The research supports the concept of organisational interdependence or cooperation rather than competition or conflict between customer and manufacturer. The traditional model which stressed market forces and the interaction of supply and demand is replaced by one which suggests that buyer and seller are not competing with each other in a situation where one inevitably gains at the other's expense. Rather, both seller and buyer gain from cooperation in new production development through the exchanges which take place. In this situation continuing organisational interdependence would appear to have replaced the traditional pattern of individual buyer–seller transactions around specific purchases (or sales). Inertia or reluctance to change supplier, caused by the lack of tangible differences in suppliers' offerings and characterised by comfortable relationships between buyer and seller, appears to have been increasingly replaced by a new form of inertia caused by the greater need for continuing relationships between buyer and seller to manage new product development and purchasing decisions effectively.

The results of this research suggest that such inertia may be professionally acceptable in product innovation and acceptance decisions. Since a stable and continuing relationship between buyer and seller was found to be necessary for successful new product development decisions, there are logical reasons to support inertia as a deliberate business strategy. This would demand a fundamental restructuring of the marketing approach adopted by some companies (for example, the British machine tool manufacturers and their customers in this survey), with greater attention paid to planning specific customer relationships over time, and less to sales of individual products.

In terms of business strategy the supplier is frequently obliged to make organisational changes to meet the terms of the contract imposed by the customer. For example, a machine tool manufacturer may be required to improve quality control checks, standards of cleanliness in assembly areas, or the general organisation which the customer will deal with. The most successful bidder in such instances is likely to be the one whose general organisation best meets the customer's requirements.

If the supplier can identify these requirements in advance, and demonstrate to the customer that the organisation can satisfy such requirements, then the probability of success will be increased. The research reported in previous chapters has identified certain general organisational characteristics which the manufacturer could usefully give attention to in advance of a specific bid to an individual customer in order to pre-empt the customer imposing such changes in the actual contract or simply buying elsewhere.

Such characteristics include the structure of the company. Here the supplier clearly needs to show to the potential buyer the various points of contact and the responsibilities of individual managers. (Many companies do this already by including organisational diagrams or 'organigrams' in their bid package.) Moreover this organisational structure will have to be flexible enough so that the supplier can convince individual customers that it meets their own requirements effectively.

General management quality was also important and the factors which appeared to facilitate interaction in earlier chapters of this research included paying greater attention to product quality and value for money through value engineering, and a willingness to keep abreast of new technologies through participation in external activities (such as collaborative ventures with universities and other research institutions, and membership of professional associations).

The final consideration which emerged from the results of this research was that the compatibility of user and supplier also appeared to depend strongly on the cultural systems within which user–supplier relationships developed. There were many aspects of West German management culture which enhanced and supported close working relationships between customer and manufacturer. These included management education with its stress on engineering and technical quality, and a greater sense of the benefits which could accrue from close collaboration. Collaboration was also significant in a general sense and the network of relationships which existed in West Germany between customer and manufacturer, manufacturer and other manufacturers, manufacturer and trade associations, manufacturer and research institutions, and manufacturer and financial institutions, was far more developed than in the United Kingdom. A future study could usefully examine the reasons for this apparently greater compatibility between the parties involved in more detail, since it clearly plays a significant role in the success or failure of new product development activities.

Implications for other sectors of the mechanical engineering industry

This study has identified several general factors which appear to inhibit the ability of British companies in the sector examined to develop new products in line with changing market conditions. Owing to their nature these factors are also likely to influence new product development activity in other related industries.

The analysis of the data points to the weakness of design in the sectors studied in the United Kingdom. This weakness is partly related to basic factors such as a failure to employ enough staff within the design function. More complex issues are also involved such as the failure to integrate design effort when companies are brought together under the financial control of a holding company. This latter factor was clearly related to an apparent reluctance to move towards modular design for many basic components used in machine tools produced within the group, and achieve economies of scale through standardisation.

Poor industry–university linkages also weakened the quality of the design effort. A concentration on theoretical rather than applied problems in the university sector appeared to be one major inhibiting factor. The linkages in West Germany were more extensive, enhancing the quality of the design effort.

A further factor weakening the quality of design was the relative compartmentalisation of functions typical in the British machine tool manufacturers studied. This meant that Design was segregated from Sales which were seen as the prime source of customer contact. As a result design staff tended to be isolated from the customer and were unable to interpret customer requirements directly. Such compartmentalisation of functions where each is restricted to specified activities has been found to be typical in other sectors of British manufacturing industry when it is compared with West Germany. There is an obvious danger that this will have the same damaging effect on the market consciousness of the design function in these sectors as it does in the machine tool industry.

Underlying the relative weakness of British product design effort is the basic problem of the relative status of technical and financial objectives in the companies studied. The pattern of ownership and control in the United Kingdom tended to stress the importance of achieving a satisfactory performance in financial terms as a member of a group of associated companies. As a result considerable emphasis was given to financial objectives at the expense of other factors such as improvements in product quality. The link between product development and long-term financial success did not appear to have been perceived by these companies who maintained a myopic view related to achieving cost savings on existing product ranges and not giving sufficient resources to new product development. Such cost savings frequently appeared to include reducing the size of the design staff. These manufacturers appeared to have adopted a defensive posture which in the long term was likely to result in the failure of the subsidiary when their financial returns were no longer acceptable in terms of overall group performance.

In contrast, in West Germany the companies' objectives appeared to stress more directly the importance of product quality as a major corporate goal. This reflected a longstanding tradition in these companies where attention was placed on producing value-for-money products which would give their customers extra benefits in their own manufacturing operation. Since such companies were more likely to be independently owned and managed, technical as well as financial goals could be used to provide direction for the company. Since the pattern of ownership and control in other sectors of the engineering industry is similarly different in West Germany and the United Kingdom, it would be anticipated that this will be an influence in the objectives of companies in these sectors too.

In terms of the customer, it is clear from the research that many of the comparative weaknesses of the suppliers in the United Kingdom are also weaknesses of the customer. In particular, the dominant influence of financial rather than technical criteria in company planning is evident. Again, patterns of ownership and control tend to stress the importance of short-run financial objectives. Investment in new plant is expected to pay back its cost rapidly, and innovation incorporated in new products or processes is discouraged because the benefits of a new technology are by definition long-term and more difficult to define. New technologies often provide benefits which are difficult to quantify using conventional appraisal techniques. For example, a new type of machine tool may provide additional benefits in terms of reliability (freedom from breakdown), reduction in materials wastage, and enhanced operator performance, which cannot enter directly into the pay-back analysis equation. New manufacturing systems have major effects not only on the quality of the finished component, but also on broader aspects of production planning and control, inventory policy, and ultimately the ability of the company to respond quickly to changes in the market with changes in its own products. These effects are often very difficult to quantify, and make the financial justification of significantly different manufacturing technologies an extremely complex process.

Whilst financial considerations were also important to the West German companies in the research, it was evident that considerable importance was given to technical performance too. West German customer organisations were more likely to be run by engineers with a keenly developed sense of maintaining an efficient manufacturing capability which would keep costs as low as possible, in a market where there was considerable emphasis on the value for money which their products provided. In their appraisal, value for money in a machine tool was seen in terms of the benefits of the machine tool (or manufacturing system) they were buying relative to the costs of that machine tool. The absolute price of the machine tool was less important than its potential net contribution. This philosophy was also apparent in their attitude to the design of their own end-product. Using efficient manufacturing equipment was one way of ensuring that their own products incorporated value for money for their customers.

This general attitude towards product quality appears to pervade other sectors of the engineering industry, and is reflected in the products manufactured by these sectors as well as the requirements of these companies as customers for plant or equipment. This is a feature of the orientation of these companies which has not arisen as a response to changing circumstances in the last twenty years or so, but rather appears to have pervaded the orientation of management for far longer. When conditions in world markets have deteriorated, it is a characteristic which provides a strong countervailing influence on the performance of these West German companies.

Since the relative importance given to technical and financial considerations is a function of a long-established pattern of management education, linked to fundamental differences in managerial goals, it would be misleading to suggest that British companies might change rapidly towards this particular orientation towards new product development. However, the lessons of this research are clear. If British companies in the sectors studied do want to make improvements then a fundamental shift in orientation is required away from a short-run fixation with lower-priced products to an orientation which stresses the importance of product quality as a major determinant of long-run international trading success.

Appendix A: The British and West German machine tool industries 1950–83

Introduction

This industry note describes the structure and recent performance of the British and West German machine tool industries, providing background material for the analysis of Chapters 4–6. The note is divided into three sections. The first of these provides a review of basic technologies in the industry, the second describes the market for metal-cutting machine tools and contrasts the relative performance of different national machine tool industries in these markets since the war. The final section examines the structure of the two national industries chosen for study.

Types of machine tool

Machine tools can be split into two basic types: those which remove metal in one way or another, and those which shape or form metal without necessarily removing any. Milling machines, lathes, and grinding machines fall into the first category, whilst presses fall into the second category. This research is predominantly focussed on metal-cutting machine tools. Within this sector there are a variety of types of machine tool, many of which have been used by industry for a considerable period of time. Machine tools themselves can frequently be relatively old. This is partly a function of the general applicability of a machine tool to certain basic functions. The removal of metal is a relatively basic task and if a machine tool satisfies this need and continues to operate, then there is no obvious reason to change it (as is the case, for example, with planing machines). The age distribution also reflects the dates when new technologies were introduced (for example, the populations of NC (numerical control) machine tools, and physico-chemical machine tools are both relatively young).

Analysis of the age of machine tools on an international basis shows that the British stock of machine tools compares favourably with machine tools in most other countries with the exception of Japan. The proportion of all machine tools under ten years old in the United Kingdom in 1979 was 39%. In West Germany it was 37%. These figures contrast markedly with the same year in Japan, where 60.5% of all machine tools were under ten years old, and 33% of machine tools were less than five years old. (Comparable figures are not available for the United States for the same year.)

Taking age as one measure of the degree of technical sophistication of a machine tool, in that younger machine tools are likely to be technically more refined than older machine tools, the figures suggest that the technical sophistication of machine tools in the United Kingdom and West Germany may be comparable, whereas those of Japan could be expected to be considerably more sophisticated. This conclusion depends not only on the relative age of the machine tool stock, but also on the 'quality' of that investment. For example, British industry may have been investing in low technology equipment, whilst West Germany may have invested in higher technology machine tools.

There have been relatively few radical changes in technology in the machine tool industry, with much of the product development effort focussed on minor innovation of the continuous variety. Major innovations (such as the introduction of numerical control) have come from developments outside the industry. In recent years these have included developments in electronics, notably computers, microprocessors and digital elements, and the development of feed drives, particularly high powered, d.c. motors.

The main areas of continuous machine tool development are research into cutting processes, spindles and bearings, machine structures, slideways, numerical and other forms of control, physico-chemical processes, and metal-forming processes. Research into cutting processes has contributed to the development of carbide, coated carbide, and ceramic tools, and to the improvement of tool geometry. These developments have been largely initiated by tool manufacturers, and not by the machine tool industry itself. Improvements in the tools have led to large increases in metal removal rates despite the growing usage of 'difficult to machine' materials. Research into spindles and bearings has produced spindle/bearing assemblies of high stiffness and power handling capacity. This has led to increased accuracy and reliability in the output of the machine tool.

The structure of the machine tool itself has also been extensively studied, and this has led to improvements in design, which permit higher rates of metal removal made possible by modern cutting tools and spindles. The trend to modular design in machine tools has also led to greater standardisation in the production of machine tools. Work on the slideways of the machine tool has led to the development of different types of slideway, new materials, and lubricants, which in turn has led to more sensitive (hence more accurate and automatic) control systems.

The areas of development described thus far are areas of continuous relatively minor change. The machine tool industry has experienced a considerable change in some technologies in the last 20 years in three further areas. The first of these is the development of numerical control technologies. Numerical control allows the operator to make the best use of skilled labour and greatly increase productivity, particularly in small batch production of relatively complex parts. Working with the original design engineer's drawing, the parts programmer can produce a control tape for a machine tool. This control tape can then be used to control the machine tool automatically through a cycle of operations with a higher degree of precision than with conventional manually operated machine tools. The machining task itself is considerably 'de-skilled'. Emphasis is placed on the programming activity which produces the tape. The benefits of numerical control in terms of improved floor-to-floor times, improvements in accuracy, and enhanced reliability have led to its increased adoption amongst machine tool users.

Further developments have focussed on refining the original numerical control process, and this work has received considerable impetus from developments in microprocessor technology. Such developments now make it possible to link together the work of a series of machine tools, and control their activities in a coordinated manner using a central computer. This has led to the concept of computer numerical control (or CNC) and direct numerical control (DNC) which manufacturers of machine tools are increasingly beginning to design for.

The concept of integrated manufacturing systems is linked to the development of 'unmanned' factories, with computer software systems providing full automation and optimisation of all steps in machining piece parts and assembly. In such a system a degree of palletisation of the piece parts is needed, and this is already apparent in the development of automatic transfer lines. A new generation of flexible manufacturing systems has now been developed which allow variations in components in the same automatic machining system. In such systems the production planner is able to vary the components and machining operations and still retain the advantages of automated production. Several of the companies visited during the course of the field work produce such systems where pallets on which the part is already loaded are mounted on tow line transporters, and taken through a system of computer-controlled machine tools, using several different types of machine tool incorporating multiple spindle head changes, washing, pallet loading and unloading.

In this instance developments in machine tools are linked with other broader developments in handling and transfer mechanisms, which will take machine tool manufacturers into new businesses. To supply such a system the manufacturer has to think in terms of the total manufacturing requirement rather than the contribution of an individual machine. This may demand a change in attitude to their business on the part of the machine tool manufacturers (this is commented on in some detail in the analysis of Chapter 4).

The development of physico-chemical processes, principally electro-deposition machining (EDM) and electro-chemical machining (ECM), also represents a major technical change. In the case of EDM a part is formed by depositing metal onto a piece by electrolysis. With ECM the part is formed by electrical erosion of the workpiece. Both technologies have proved to be important steps forward. EDM now makes a major contribution to productivity because in many instances it is more rapid than conventional machining technologies, and ECM enables the manufacture of components in materials that could not be economically machined in any other way. Finally, other metal-forming processes are being developed which produce metal components to increasingly higher degrees of precision without the need for machining. For some applications such processes may ultimately replace the metal-cutting machine tool.

The acceptance of technical change is ultimately influenced by economic factors, which have had a major role in determining the way in which new technologies have developed. Of these the major factor has been a rise in the cost of skilled labour, which has forced manufacturers to adopt new (capital-intensive) technologies. The absolute scarcity of labour has also been an important consideration.

As machine tools have become more complex and have become more productive (essentially de-skilling many machining operations), they have also become more expensive. This has led to requirements for increased reliability and easier maintenance, in order that the machines can be available for use as much as possible. Emphasis has been needed in design on increasing utilisation, particularly minimisation of time lost in setting up, in tool changing, loading and unloading of the workpiece, and the inspection of dimensions.

The market for machine tools

The demand for machine tools is characterised by considerable fluctuations on a year-to-year basis, typical of an industry where the demand for its own products depends on (or is derived from) the demand for other products. Recessions in the industry tend to last longer than recessions in industry as a whole, and periods of expansion and high demand for its products tend to be shorter than those of industry as a whole. In recent years periods of recession have led to a considerable reduction in the number of people employed in the British and West German machine tool industries, and periods of expansion have been accompanied by complaints that machine tool manufacturers in the United Kingdom cannot expand output as rapidly as overseas competitors can to exploit the temporary market expansion.

Table A.1 compares market shares of the top ten machine tool manufacturing countries in 1977 and 1981.

In 1977 West Germany was the world's leading producer of machine tools reflecting the reputation which its products had gained in the years since the second world war. This level of performance had been sustained by a high level of exports which supported domestic production. Between 1977 and 1981 West Germany lost this lead to the United States. The most significant

Table A.1. *Shares of world machine tool production 1977 and 1981*

Country	Market share 1977 (%)	1981 (%)
West Germany	17	15
United States	16	20
Soviet Union	15	12
Japan	10	18
Italy	6	5
East Germany	5	3
United Kingdom	5	3
France	4	3
Switzerland	4	3
Poland	4	—
Romania	—	2

Source: *American Machinist* 1978 and 1981.

development, however, was the rapid growth of the Japanese machine tool industry during the same period (from 10% of world machine tool production in 1977 to 18% in 1981). British companies continued to experience a decline in output falling from 5% of world production in 1977 to 3% in 1981.

Table A.2 analyses world machine tool trade balances for the top ten producing countries in 1977. The table shows the heavy dependency of some countries (notably West Germany, Switzerland and East Germany) on the export market. Japan and the United States on the other hand appear to rely more upon higher levels of domestic consumption. Since 1977 the picture has changed for Japanese companies, which now export far more of their output.

The British market for machine tools

Machine tools are used in virtually the whole of manufacturing industry, although they are concentrated in some key sectors such as the car industry, office machinery, general mechanical engineering and metal goods industries.

The machine tool industry is one of the earliest to feel the effect

Table A.2. *World machine tool trade balances for the top ten countries (value in US million dollars)*

Country		1977 Exports	Imports	Balance	Exports as % of sales
1. West Germany	(1)	1,911.8	289.2 +	1,622.6	73%
2. Japan	(3)	566.4	89.4 +	477.0	36%
3. Switzerland	(2)	474.9	73.3 +	401.6	84%
4.* East Germany	(4)	535.0	205.0 +	330.0	76%
5. Italy	(6)	408.1	181.4 +	226.7	48%
6.* Czechoslovakia	(7)	197.1	114.9 +	82.2	64%
7. United Kingdom	(8)	324.4	245.5 +	79.9	46%
8. United States	(5)	446.9	400.3 +	46.6	19%
9.* Taiwan	(–)	50.5	30.0 +	20.0	n.a.
10.* Belgium	(10)	110.8	93.7 +	17.1	n.a.
World totals		6,321.7	6321.7	—	—

*Based on available fragmentary data. 1976 placing in brackets.
Source: *American Machinist*, February 1978.

of a depression since the key industries always cut back first of all on their expenditure on new plant and machinery. On the other hand, in a period of boom these same industries have tended to invest heavily in machine tools which causes a sharp surge in machine tool orders and a traditional period of glut and delayed delivery time following the period of famine. Demand for machine tools has tended in the past to fluctuate in a fairly regular cyclical pattern.

This feature was highlighted in the report of the 'Bullock' Committee in 1966. Peaks occurred in 1958 and 1960 and troughs in 1959 and 1963. However, the committee emphasised that a strong competitive machine tool industry should be able to rise above this cyclical performance, when it stated:

> there can indeed be no certainty that any moderation of the cyclical pattern which it may be possible to achieve will of itself lead to a significant strengthening of the machine tool industry: on the other hand there can be no doubt that the stronger, more virile and the more competitive the indus-

> try, the less will be its preoccupation with the cycle and the greater its ability to rise above it.

Since the Bullock Committee's report the overall position of the British machine tool industry has worsened rather than improved. The trend in orders has continued to decline, and as a consequence cyclical problems have increased. Regular cyclical downturns in machine tool orders in the United Kingdom occurred in 1968, 1971, 1975, and 1979 and a further trough in orders has now developed in line with the general deflation of the world economy. The depressions in the 1970s and 1980s have shown a tendency to last longer and be deeper than in the 1960s. Rather than the periods of boom lasting for about three years with a short depression of about a year, the position is now almost reversed with the troughs lasting longer and the subsequent boom only lasting for a short time.

This problem was highlighted in a report on the machine tool industry commissioned by the Joint Shop Stewards' Committee of T.I. Churchill Limited in March 1978. The North East Trade Union Studies Information Unit which produced this report concluded that the industry was facing significant long-term problems, not necessarily connected with the downturn in the current trade cycle. One section of this report commented: 'The implication . . . is that these may be "more normal times" than we imagine, and that the present over-capacity is not merely due to cyclical problems, but also due to a substantial permanent reduction in the United Kingdom's real order book and future prospects.'

As well as the declining trend in output and production since 1970, there has been a rise in the imports of machine tools entering the United Kingdom with a few notable countries making rapid penetration into the domestic market in this period. Import penetration in the British market increased from 18% of domestic consumption in 1975 to 40% in 1981. Table A.3 illustrates the United Kingdom's top 12 leading machine tool import sources. From Table A.3 it can be seen that between 1975 and 1982 West Germany has consistently been the principal source of supply of new machine tools to the United Kingdom. The most marked change has been the increase in the share which Japanese companies have of British imports (rising from 4% in 1975 to 11% in 1982).

Import penetration was recognised as a major threat to the British machine tool industry even in the period when the market

Table A.3. *The United Kingdom's top ten leading machine tool import sources (values in £ million)*

	1975	1982
West Germany	34%	32%
United States	13%	14%
Switzerland	10%	8%
France	7%	5%
Italy	6%	8%
Japan	4%	11%
Sweden	4%	—
Netherlands	3%	2%
Spain	3%	—
East Germany	2%	—
Belgium	—	1%
Irish Republic	—	—
Taiwan	—	—
World	100% (£114.7 million)	100% (£133.0 million)

Source: *Overseas Trade Statistics for the United Kingdom for 1982*, HM Customs and Excise 1983.

for machine tools remained comparatively buoyant. In 1965 a NEDO survey of British machine tool users were asked their reasons for increasingly opting for imported machines: 5% said price was the main factor, 5% better aftersales service, 8% the willingness of foreign manufacturers to meet special requirements, 11% the prospect of reciprocal trading agreements, 20% quick and reliable delivery, 21% the machine specification not being available in Britain, and 30% the technical superiority of the foreign machine. If all the technical factors are combined then the technical superiority of the imported product was given as the main reason for choice by 59% of the British customers. Price appeared to be a less important factor.

In terms of world trade the British machine tool industry held an 8% share of world production from 1966 to 1971. This had fallen to 6% by 1975 and 3% by 1981. The major export markets for British machine tools are listed in Table A.4. The predominant importance of the United States as a customer country is clear in this table.

Table A.4. *The United Kingdom's top 12 leading machine tool export markets (values in £ million)*

1975		1981	
United States	17.8	United States	25.7
South Africa	13.5	West Germany	8.2
Poland	11.1	South Africa	5.1
France	8.6	Canada	2.7
Spain	8.4	Soviet Union	2.3
West Germany	8.2	India	1.5
Brazil	7.0	France	1.2
Sweden	6.8	Argentina	1.2
Australia	6.3	Switzerland	0.9
Soviet Union	5.9	Irish Republic	0.7
Canada	5.7	Mexico	0.4
Italy	5.0	Australia	0.2

Source: *Overseas Trade Statistics of the United Kingdom 1981*, HM Customs and Excise 1982.

The structure of the British machine tool industry

In 1981 approximately 150 companies were engaged exclusively in the manufacture of machine tools in the United Kingdom. Although many of these companies were members of large groups they tended to have retained their identities as specialist manufacturers of particular types of machine tool. The 1960s and early 1970s marked a major period of restructuring within the British machine tool industry which led ultimately to the formation of major groupings of machine tool manufacturers. Table A.5 lists the major companies wholly or partially engaged in machine tool manufacture in 1981.

Measures of market concentration are difficult to provide since the turnover of machine tool manufacturers is often not disclosed separately in group accounts. However, the market is clearly dominated by the major groups of machine tool manufacturers. (For example, one estimate made by the author in 1980 identified seven major companies or groups accounting for 60% of the output of the industry.) There are also smaller companies operating successfully on the fringes of the market.

One notable feature of the industry is the extent of American

Table A.5. *Major British companies wholly or partially engaged in machine tool manufacture 1981*

Company	Turnover 1981 (£ million)
Wickman	63.9
Cincinnati Milacron	47.5
Butler Elliott	111.9
Staveley Machine Tools	16.4
600 Group	148.5
Kearney & Trecker Marwin	24.6*
Wadkin Ltd	27.1*
Jones & Shipman Ltd	16.67*
T.I. Machine Tools Ltd	n.a.
Adcock & Shipley Ltd	139.4*

*1980 figure.

participation. Cincinnati Milacron Ltd and Adcock and Shipley Ltd are both owned by American parent companies. In this case ownership is associated with extensive control and support of product development and general marketing policy.

Cincinnati Milacron produce a substantial proportion of the United Kingdom's output of NC machine tools. When the author contacted this company to arrange a visit, he was told that there would be little point in discussing product development work in the British company, since this was done centrally in the United States with minimal participation from the United Kingdom. Adcock and Shipley is wholly owned by Textron Inc., and in addition to manufacturing its own range of products also manufactures the Bridgeport range from designs produced by the parent American company. The Bridgeport series of products has proved very successful in markets worldwide.

A second notable feature of the industry which is evident from the analysis of case study material is that many British manufacturers act as agents for overseas suppliers (for example, Staveley Machine Tools and T.I. Machine Tools Divisions). British companies were far more likely to act as agents for overseas manufacturers than their competition in West Germany.

Reflecting the restructuring of the industry in the 1960s and 1970s, the number of employees working in the machine tool indus-

Table A.6. *Changes in the numbers employed in the UK machine tool industry 1960–82*

1960	1964	1968	1972	1976	1980	1982
81,000	83,000	86,000	68,000	62,000	56,000	44,000

Source: *Employment Gazette*: Department of Employment 1983 (figures refer to minimum list heading 332 of the standard industrial classification which includes metal-cutting machine tools, metal-forming machine tools, welding and flame cutting equipment and other physico-chemical process cutting and forming equipment).

try has declined considerably. Table A.6 illustrates the dramatic changes in the numbers employed in the British machine tool industry 1960–82.

The 'Way' report (1970) viewed this concentration as essential in ensuring the continued survival of a few key British machine tool manufacturers. The authors commented:

> Scope exists for product rationalisation and concentration because –
> (a) The cost of developing, manufacturing and selling advanced machines and systems of production is such that a high market share is needed to support the effort needed.
> (b) Competition in the conventional general purpose machine tool field is such that large scale production is necessary if the British Industry is to remain a viable supplier at home and abroad.

However, whilst concentration and large-scale manufacturing may reduce the unit costs of standard products, it is not in itself conducive to the production of special purpose technically sophisticated products built to customers' individual needs. The industry's present structure therefore may well be one major barrier to the development of new products.

The West German machine tool industry

Export and import performance

As Table A.1 indicates the West German machine tool industry is one of the largest suppliers of machine tools on a worldwide basis.

Its share of world machine tool exports varied around 35% throughout the 1970s, whilst by comparison British machine tool manufacturers consistently lost market share. More recently, however, its position in many export markets has been threatened by the rapid growth of the Japanese machine tool industry.

Imports have consistently increased from approximately 20% of domestic consumption in 1960 to 30% in 1978, but the rate of increase is not as marked as that experienced in other countries. During the same period imports as a percentage of domestic consumption rose from 5% to 21% in the United States, and 21% to 47% in the United Kingdom. By comparison in Japan the same period imports fell from 31% to 8%, reflecting an increased reliance on domestic production.

In West Germany machine tool production has consistently been aimed at the entire world market rather than oriented to home needs (the capacity of the industry is considerably greater than domestic requirements). This global approach has allowed the West German manufacturer to switch volume when demand in one country falls away, and also allows production runs long enough to achieve economies of scale. At the same time a considerable amount of effort has been put into improving the technology of the product to counter the effects of the continuing strength of the Deutschmark against other currencies.

More than 30% of West German machine tool exports in 1978 went to Eastern European countries, and the machine tools sold tended to be relatively less sophisticated than those demanded by other markets. However, this business has been won at a cost. Many East European buyers insist on barter or part-barter deals, which have frequently forced West German manufacturers to accept in part payment East European equipment which they are then obliged to sell in West Germany.

Structure

In 1978 approximately 350 companies in West Germany concentrated solely on the manufacture of machine tools. Twelve of these companies employed over 1000 people, 33 employed between 501 and 1000 and 65 employed between 250 and 500. Contrasting with the United Kingdom much of the production of machine tools is concentrated in medium sized independent companies frequently owned by the family of the original entrepreneur who founded the company (50% of the output of the industry in 1978 was produced by companies employing less than 500 people). Few of these companies are publicly quoted.

Some commentators have argued that this structure gives the West German industry considerable advantages. For example, K. Gooding (1978) commented:

> In the past the West German accounting practice which allows manufacturing companies to build up a decent layer of financial fat to see them through the lean periods of the demand cycle has helped the machine tool makers survive. Couple this with the immense flexibility of the German industry, made up as it is in the main of medium sized family companies able to make swift decisions, and act on them immediately, and it is not difficult to understand why some of its European neighbours have seen West Germany as the major threat to their home grown manufactures.

Despite the apparent strengths of the West German industry it did not escape the recession in the market in the 1970s. Between 1973 and 1977 employment in the West German machine tool industry fell by 13% from 111,500 to 97,000. Since then this employment has continued to fall. However, as earlier sections of this note indicate, the industry maintained its share of world markets during this period.

Appendix B: Previous research evidence

Table B.1. *Factors influencing organisational adoption behaviour: previous research evidence*

Researcher	Profitability of firm	Size of investment	Durability of existing investment	Expanding output	Time	Phase of business cycle
Abu-Ismail (1976)						
Baker (1971)	+ S					
Baker & Parkinson (1976)	+	+	−	+	+	
Bernhardt (1970)						
El Sherbeny (1978)						
Gold, Peirce & Rosseger (1970)				+		
Mansfield (1963*a*)	+ S	− S	− NS	+ NS	+ NS	+ NS
Mansfield (1963*b*)	+ NS		− NS			
Mansfield (1972)				+ S		
Martino, Chen & Lenz (1978)	+ NS		− NS			+ NS
Metcalfe (1972)						
OECD (1970)						
Peterson, Rudelius & Wood (1972)						
Ray (1969)						
Webster (1971)						

	Sensitivity of demand for product it produces to a reduction in price or improvement in quality	Relative factor prices	Sales volume	Degree of vertical integration	Risk propensity	Capital cost of machine	Industry growth rate
Abu-Ismail (1976)	+ S						
Baker (1971)							
Baker & Parkinson (1976)	+		+			–	
Bernhardt (1970)							
El Sherbeny (1978)	+						
Gold, Peirce & Rosseger (1970)							
Mansfield (1963*a*)							
Mansfield (1963*b*)					– S		
Mansfield (1972)							
Martino, Chen & Lenz (1978)						– NS	
Metcalfe (1972)				+			
OECD (1970	+	+					
Peterson, Rudelius & Wood (1972)					+ S		
Ray (1969)							
Webster (1972)					+		

Table B.1 (*cont.*)

	Size of firm (assets)	Size of firm (employees)	Expected return from investment	Growth rate	Liquidity of firm	Market concentration	Number a firm has to buy to go from 10% to 90% adoption	Compatibility of investment
Abu-Ismail (1976)	+ S		+ S					
Baker (1971)	+ S	+ S						
Baker & Parkinson (1976)	+	+	+					
Bernhardt (1970)			+ S					
El Sherbeny (1978)	+		+					
Gold, Peirce & Rosseger (1970)								
Mansfield (1963*a*)								
Mansfield (1963)	+ S		+ S	+ NS	+ NS	+ NS	− NS	
Mansfield (1972)	+ S							
Martino, Chen & Lenz (1978)			+ NS					+ S
Metcalfe (1972)	+	+						
OECD (1970)			+					
Peterson, Rudelius & Wood (1972)	+ S	+ S		+ S				
Ray (1969)			+		+			
Webster (1971)	+	+	+		+			

+The researcher hypothesised a positive relationship between the factor and the speed with which an organisation adopted an innovation.

−The researcher hypothesised a negative relationship between the factor and the speed with which an organisation adopted an innovation.

NS The relationship when tested was shown to be statistically significant.

NS The relationship when tested was shown to be not statistically significant.

Where neither S nor NS appear the relationship was not statistically tested.

Table B.2. *Attributes of significance in evaluation of innovation: evidence from a survey of British manufacturing industries*

	Degree of importance			Total response		
Product attribute	Very important 1 + 2 + 3	4 + 5 + 6	Very important 7 + 8 + 9	$\bar{x}$	n	N
Quality of output	89% (587)	7% (49)	4% (27)	1.9	100% (663)	704
Number of uses	33% (168)	35% (177)	32% (162)	4.9	100% (507)	700
Lower price and product	63% (375)	23% (141)	14% (81)	3.2 (5)	100% (597)	675
Higher profit margins	73% (447)	20% (126)	7% (43)	2.7	100% (616)	682
Skilled labour savings	67% (347)	15% (80)	18% (93)	3.2	100% (520)	687
Semi-skilled labour savings	52% (264)	30% (154)	17% (88)	3.8	100% (506)	692
Unskilled labour savings	37% (141)	24% (91)	39% (151)	5.1	100% (383)	681
Raw material savings	57% (239)	19% (81)	24% (100)	3.8	100% (420)	689
Lower running costs	52% (343)	28% (186)	21% (137)	3.9	100% (666)	697
Less floor space	25% (156)	31% (194)	44% (277)	5.6	100% (627)	702
Long life	44% (299)	42% (282)	14% (92)	4.1	100% (673)	699
Technical specification	81% (539)	13% (90)	6% (38)	2.3	100% (667)	697
Reliability	94% (652)	4% (27)	2% (13)	1.6	100% (692)	700
Noise	41% (243)	37% (218)	23% (135)	4.4	100% (596)	698
Cleanliness	48% (288)	34% (203)	18% (109)	4.0	100% (600)	697
Safety	69% (512)	16% (104)	5% (34)	2.3	100% (650)	701
Reduction in downtime	74% (423)	17% (97)	8% (48)	2.7	100% (568)	693
Return on investment	85% (581)	11% (73)	4% (26)	2.0	100% (680)	703

Notes to Table B.2

Ranks 1, 2 and 3, 4, 5 and 6, 7, 8 and 9 have been grouped together for purposes of comparison. Hence 587 in the 1 + 2 + 3 column, 'quality of output' row indicates that 587 respondents rated quality of output as either 1, 2 or 3 on the importance scale.

$\bar{x}$ is the mean rating of the product attribute, which has been calculated under the assumption that the points on the scale are equidistant (i.e. that the scale has interval properties).

n is the number of respondents who answered by ringing a reply between 1 and 9 on the scale, whereas N is the total number of respondents who replied to the question. Consequently $N - n$ is the number of respondents who indicated that the attribute was not applicable (not shown in table), and $706 - N$ is the number of respondents who failed to answer the question (again not indicated). (706 equals the number of questionnaires used in the analysis.)

Reference: Baker and Parkinson (1976).

References

Chapter 1

British Institute of Management (1978), 'Spur Initiative', *Field Survey Reports – Key Problem Areas in Engineering*, October 1978.

National Economic Development Committee (1981), Internal Report.

NEDO (1977), *International Price Competitiveness and Export Performance – the Non-Price Factors*, NEDO, London, April 1977.

Pavitt, K. (ed.) (1980), *Technical Innovation and British Economic Performance*, Macmillan, London.

Phelps-Brown, H. (1977), 'What is the British Predicament?', *Three Banks Review*, December 1977.

Rothwell, R. (1979), 'The Relationship between Technical Change and Economic Performance in Mechanical Engineering: Some Evidence', in *Industrial Innovation: Technology, Policy Diffusion*, ed. M.J. Baker, Macmillan, London.

Chapter 2

Abu-Ismail, F.A.F. (1976), 'Predicting the Adoption and Diffusion of Industrial Product Innovations', unpublished Ph.D. Dissertation, University of Strathclyde.

Aiken, M. and Hage, J. (1968), 'Organisational Inter-Dependence and Intra-Organisational Structure', *American Sociological Review*, vol. 33, pp. 912–30.

Baker, M.J. (1975), 'The Diffusion of Industrial Innovation: An Exploration of Factors Associated with the Process, their Measurement and Predictive Utility', D.B.A. Dissertation, Harvard University 1971, published in Baker, M.J., *Marketing New Industrial Products*, Macmillan, London.

Baker, M.J. and McTavish, R. (1976), *Product Policy and Management*, Macmillan, London.

References

Baker, M.J. and Parkinson, S.T. (1976), 'Predicting the Adoption and Diffusion of Industrial Innovation', final report to SSRC on research grant.

Burns, T. and Stalker, G. (1961), *The Management of Innovation*, Quadrangle Books, Chicago.

Carter, C.F. and Williams, B.R. (1957), *Industry and Technical Progress*, Oxford University Press, London.

Cooper, R.G. (1980), 'Project New Prod: Factors in New Product Success', *European Journal of Marketing*, vol. 14, nos. 5/6.

El Sherbeny, A.F.M. (1978), 'Behavioural and Organisational Influences on the Adoption of Industrial Product Innovation', unpublished Ph.D. Dissertation, University of Strathclyde.

Gronhaug, K. (1976), ' "Variation" as an Innovative Strategy in New Product Development', *Industrial Marketing Management*, vol. 5.

Hakansson, H. (ed.) (1982), *International Marketing and Purchasing of Industrial Goods*, John Wiley & Sons, Chichester.

Hakansson, H. and Ostberg, C. (1975), 'Industrial Marketing: An Organisational Problem', *Industrial Marketing Management*, vol. 4, pp. 113–23.

Johne, A. (1982), 'Innovation, Organisation and the Marketing of High Technology Products', Ph.D. Dissertation, University of Strathclyde, Department of Marketing.

Mansfield, E. (1963), 'Speed of Response of Firms to New Techniques'. *Quarterly Journal of Economics*, May 1963, pp. 290–312.

Mattson, L.G. (1975), 'Impact of Stability in Supplier–Buyer Relations of Innovative Behaviour in Industrial Markets', paper discussed at symposium 'New Dimensions in Marketing', Voksenasen, Oslo, 2–5 April 1975.

Parkinson, S.T. (1975), 'The Role of Information in the Adoption of Industrial Innovation', unpublished M.Sc. Dissertation, University of Strathclyde, Department of Marketing.

Parkinson, S.T. (1980), 'User–Supplier Interaction in New Product Development', Ph.D. Dissertation, University of Strathclyde, Department of Marketing.

Parkinson, S.T. (1982), 'Successful New Product Development: Having a Good Customer Helps', *The Business Graduate*, Special Issue – *Innovation*, Spring 1982.

Pessemier, E.A. (1966), *New Product Decisions*, McGraw-Hill, New York.

Pugh, D.S., Hickson, D.J., Hinings, C.R. and Turner, C. (1968), 'Dimensions of Organisation Structure', *Administrative Science Quarterly*, vol. 13, pp. 65–105.

Pugh, D.S., Hickson, D.J., Hinings, C.R. and Turner, C. (1969), 'The

Context of Organisation Structures', *Administrative Science Quarterly*, vol. 14, pp. 91–114.

Robertson, A.S. (1977), 'Technological Innovation and the Management of R and D', in *Management Bibliographies and Reviews*, ed. D. Ashton, MCB Publications, Bradford.

Robertson, A. and Fox, M. (1979), 'A study in "Real Time" of the Innovation Process in Two Science Based Companies', in *Industrial Innovation: Technology, Policy, Diffusion*, ed. M.J. Baker, Macmillan, London.

Twiss, B. (1974), *Managing Technological Innovation*, Pergamon Press, London.

Von Hippel, E. (1977), 'Transferring Process Equipment Innovations from User-Innovators to Equipment Manufacturing Firms', *R and D Management*, October 1977.

Chapter 3

NEDO (1977), *International Price Competitiveness and Export Performance – the Non-Price Factors*, NEDO, London, April 1977.

Chapter 4

Gooding, K. (1978), 'Machine Tool Survey', *Financial Times*, 27 November 1978.

Joint Shop Stewards Committee, T.I. Churchill Ltd (1978), Machine Tool Report, produced by North East Trade Union Studies Information Unit, March 1978.

Lawrence, P. (1980), *Managers in Management in West Germany*, Croom Helm, London.

Way Committee (1970) – Machine Tool Export Committee under the Chairmanship of Sir Richard Way, KCB, CBE, *The Machine Tool Industry*, HMSO, London.

Chapter 5

Lawrence, P. (1980), *Managers in Management in West Germany*, Croom Helm, London.

NEDO (1965), *Survey of Investment in Machine Tools*, HMSO, London.

Appendix A

Bullock Committee (1966) – Ministry of Technology Working Party under the Chairmanship of R.H.W. Bullock, *Report of the Working Party on the Problems Arising from the Cyclical Pattern of Machine Tool Orders*, HMSO, London.

Gooding, K. (1978), 'Machine Tool Survey', *Financial Times*, 27 November 1978.

Joint Shop Stewards Committee, T.I. Churchill Ltd (1978), Machine Tool Report, North East Trade Union Studies Information Unit, March 1978.

NEDO (1965), *Survey of Investment in Machine Tools*, HMSO, London.

Way Committee (1970) – Machine Tool Export Committee under the Chairmanship of Sir Richard Way, KCB, CBE, *The Machine Tool Industry*, HMSO, London.

Appendix B

Abu-Ismail, F.A.F. (1976), 'Predicting the Adoption and Diffusion of Industrial Product Innovation', unpublished Ph.D. thesis, University of Strathclyde.

Baker, M.J. (1971), 'The Adoption of Industrial Products', *Marketing Science Institute Working Paper*, September 1971.

Baker, M.J. and Parkinson, S.T. (1976), 'Predicting the Adoption and Diffusion of Industrial Innovation', final report to SSRC, on research grant.

Bernhardt, I. (1970), 'Diffusion of Catalytic Techniques through a Population of Medium Sized Petroleum Refining Firms', *Journal of Industrial Economics*, vol. 19, no. 1, november 1970, pp. 50–65.

El Sherbeny, A.F.M. (1978), 'Behavioural and Organisational Influences on the Adoption of Industrial Product Innovation', unpublished Ph.D. Dissertation, University of Strathclyde.

Gold, B., Peirce, W.S. and Rosseger, G. (1970), 'Diffusion of Major Technological Innovations in US Iron and Steel Manufacturing', *Journal of Industrial Economics*, July 1970.

Mansfield, E. (1963*a*), 'Speed of Response of Firms to New Techniques', *Quarterly Journal of Economics*, May 1963, pp. 290–312.

Mansfield, E. (1963*b*), 'Intra Firm Rates of Diffusion of an Innovation', *Review of Economics and Statistics*, November 1963.

Mansfield, E. (1972), *Research and Innovation in the Modern Corporation*, Macmillan, London.

Martino, J.P., Chen, K. and Lenz, R.C. (1978), 'Predicting the Diffusion Rate of Industrial Innovation', final report on grant PRA76-17188, National Service Foundation, University of Dayton Research Institute, Dayton, Ohio, March 1978.

Metcalfe, J.S. (1972), 'Diffusion of Innovation in the Lancashire Textile Industry', Manchester School of Economic and Social Studies.

References

OECD (1970), 'Gaps in Technology', Comparisons between Member Countries in Education, Research and Development, Technological Innovation, International Economic Exchanges, Paris.

Peterson, R.A., Rudelius, W. and Wood, G.L. (1972), 'Spread of Marketing Innovation in a Service Industry', *Journal of Business*, October 1972.

Ray, G.F. (1969), 'The Diffusion of New Technology: Ten Processes in Nine Industries', *NIER*, vol. 48, May 1969.

Webster, F.E. (1971), 'Communication and Diffusion Processes in Industrial Markets', *European Journal of Marketing*, vol. 5, no. 4.

Index

Index

Index

British m/c tool industry – roderick floud.